the **imperfect**
environmentalist

a practical guide to
clearing your body,
detoxing your home,
and saving the earth
(without losing
your mind)

Sara
Gilbert

the imperfect
environmentalist

Ballantine Books Trade Paperbacks ⬤ New York

A Ballantine Books Trade Paperback Original

Copyright © 2013 by Gilby Inc.

Published in the United States by Ballantine Books, an imprint of The Random House Publishing Group, a division of Random House, Inc., New York.

BALLANTINE and the HOUSE colophon are registered trademarks of Random House, Inc.

Library of Congress Cataloging-in-Publication Data

Gilbert, Sara.
 The imperfect environmentalist : a practical guide to clearing your body, detoxing your home, and saving the earth (without losing your mind) / Sara Gilbert.
  p. cm
Includes index.
 ISBN 978-0-345-53758-4 (pbk. : alk. paper)
 eBook ISBN 978-0-345-53759-1 (ebook)
 1. Sustainable living. 2. Environmentalism. I. Title.
 GE196.G55 2013
 640.28'6–dc23   2012046541

Printed in the United States of America on 100% post-consumer recycled stock

www.ballantinebooks.com

9 8 7 6 5 4 3 2 1

First Edition

To my children, Levi and Sawyer, who make me want to be a better person and make the world a better place.

contents

introduction   xiii

one : clean eating and drinking   3

We Are What We Eat ■ Even the Water's Not Safe? ■ Organic Matters: You Are What You Eat ■ Eating Local: Because You Wouldn't Drive All Day to Pick Up a Tomato ■ Eating Out: From Greasy Spoons to Fine Green Dining ■ Where Does Your Meat Come From? Do You Even Want to Know? ■ The Drive-thru: Going Nowhere Fast ■ *Who Knew? Climate Change* ■ Farmed Fish, Dyed Fish, Red Fish, Blue Fish ■ Love Animals, Don't Eat Them ■ Hug Me, I'm Vegan ■ The Raw Deal: Why Bother Cooking? ■ Macrobiotics: Extreme Balance ■ Soy Happy ■ *Who Knew? The Return of the Salmon* ■ Grocery Shopping: Beyond Paper or Plastic ■ GMOs: What Did They Just Do to That Plant? ■ Juice This ■ Caffeinate Me: All About Coffee ■ Tea Time ■ The Jolly Green Drinker: Beer, Wine, and Spirits ■ Chocolate Without the Child Labor, Please ■ Sugar Isn't So Sweet ■ Nontoxic Food Preparation: Would You Eat Your Cutting Board?

two : clean house   29

Your Own Personal Green House ■ From Tree Houses to McMansions: How Much Space Do You Need? ■ *Who Knew? Habitat Destruction* ■ Home Inspections: What to Ask ■ Green Design and Construction ■ Hitting the Floor ■ Raising the Roof ■ Out the Window: Windows and Weatherizing ■ Living Off the Grid ■ *Who Knew? "The Garbage Warrior"* ■ Warming Up and Cooling Off ■ Humidifiers and Dehumidifiers: Boring but Necessary ■ Renovation: Things to Watch Out For Besides the Budget ■ Asbestos: Remedying a Bad Building Idea ■ Solar Power: Because the Oil Companies Don't Own the Sun ■ Wind Power: Blow on This ■ Lead and Other Reasons Not to Lick the Walls ■ Jump In: Nontoxic Pools ■ Clearing the Deck

three : clean home   49

Change Begins at Home ▪ Buying Furniture: Why Your Couch Is the Enemy ▪ Sleep on It: Bedding ▪ Can't I Just Wash the Toxins Off with Soap or Sanitizer? ▪ Paint It Black ▪ *Who Knew?* Greenwashing ▪ Pest Control: Because Sometimes Asking Mice to Leave Nicely Doesn't Work ▪ Bathrooms: Money Like Water ▪ Don't Flush That Pill: Pharmaceuticals in Wastewater ▪ *Who Knew?* Frickin' Fracking ▪ Hippie Household Cleansers ▪ Dishing It Out: Your Dishwasher Just Killed That Fish ▪ Cleaning the Furniture: Sit on This ▪ Books, Newspapers, and Magazines: Worth the Dead Trees They're Printed On? ▪ Dog and Cat Care: Because Fido and Mittens Want to Be Green, Too ▪ Dirty Laundry: The One-Page Version ▪ Clean Linens: The Cheat Sheet ▪ Televisions: Yes, This Hippie Has One ▪ *Who Knew?* Geothermal and Hydroelectricity ▪ Computer Geek ▪ Call Me: Cell Phones vs. Landlines ▪ Recycle Everything but Your Bad Jokes ▪ Upcycling vs. Downcycling ▪ All Plastics Are Not Created Equal ▪ Happy Green Holidays

four : clean garden   75

I Don't Garden, but You Should ▪ Composting: Sifting Through It ▪ Organic Gardening: Getting Dirty ▪ *Who Knew?* What Happened to the Honey Bees? ▪ Container Gardening: Because You Don't Need a Yard to Grow Things ▪ Fertilizers and Other Nice Names for Guano ▪ Pesticides: What Keeps Away Insects Poisons You, Too ▪ Rain Catchment: I Mean, Shouldn't Water Be Free? ▪ *Who Knew?* Dirt Worshipper ▪ Raising Chickens ▪ Goats Are the New Cow ▪ Urban Farming: Let Someone Else Do the Digging

five : clean health and beauty   89

Pretty Shouldn't Be Poisonous ▪ Lipstick Hugger: Friendly Cosmetics ▪ Hair Care: Green Locks Aren't Just for Punk Rockers ▪ Tooth Care: Chew on This ▪ Skin Deep: Your Moisturizer Just Got in My Bloodstream ▪ Paint It Red: Manicures and Pedicures ▪ *Who Knew?* Pretend You're a River ▪ Clothing: Because Even Hippies Shouldn't Have to Go Naked All the Time ▪ Yoga: Just Say Aum ▪ Tai Chi and Qigong: Not Just for the Freaks in the Park ▪ Take a Hike: Exercising Outside ▪ *Who Knew?* The Green Gym Powers Itself ▪ Sneezing Is Funny, but Allergies Aren't ▪ Acupuncture: Poking Fun ▪ Your Herbal First Aid Kit ▪ Take a Pill: Vitamins and Supplements ▪ *Who Knew?* The Amazon Rainforest: Queen of the Weather System ▪ If You Must Smoke, Don't Smoke Rat Poison

six : clean community   107

We're All in This Together ▪ Plant a Tree ▪ *Who Knew?* Clean Air, Clean Life ▪ Thrifting: Not Just for When You're Broke ▪ Free Stores: Where Everything Is, Yes, *Free* ▪ Community Gardens: Grow Food, Meet Babes ▪ Policy Wonk: Affecting Change in Local Government ▪ Getting Involved in Consumer Campaigns ▪ A Broad Church: Environmentalism and Faith Communities ▪ If You Lived Here You'd Be Home by Now: Small Growth Projects ▪ Change Starts at Home ▪ *Who Knew?* Green Eggs and Ham Isn't Greener Than Greensburg

seven : clean work and money   121

Be That Annoying Person at Work ▪ Buy Less: The Importance of Being Cheap ▪ Environmental Careers: Save the Earth, Make Money ▪ Green Grants: They'll Pay Me for That? ▪ *Who Knew?* Water Over the Dam ▪ When Your Company Is Destroying the Planet ▪ The Big Commute ▪ The Eco Office: Paper Is Over-rated ▪ Sweatshops: The Real Price Tag ▪ *Who Knew?* Nuclear Power and Nuclear Waste ▪ Investing: Banking on a Sustainable Future

eight : clean transportation
and travel   133

Getting There Is Half the Fun ▪ Cars: A Necessary Evil? ▪ *Who Knew?* Your Carbon Footprint ▪ Learn to Drive: Don't Floor It Like SpongeBob ▪ Biofuels: Or How to Run Your Car on Old Restaurant Grease ▪ Take Your Gas and Shove It: Fuel Cells and Electric Cars ▪ Public Transportation: Get on the Bus ▪ Motorcycles and Scooters: Born to Ride ▪ Bicycles: Put the Fun Between Your Legs ▪ Walk It Off ▪ Air Travel: Your Carbon Footprint on Steroids ▪ *Who Knew?* The Forest Through the Trees ▪ Geotourism: Seeing the World Without Wrecking It

nine : clean parenting   147

I Know It's Corny to Say "The Children Are the Future," but, Hey, They Are ▪ Seven Generations: Think About Your Kids' Kids ▪ Pregnancy: Your Belly Is the Perfect Eco Apartment ▪ Adoption: Because There Are Plenty of People Already Here ▪ *Who Knew?* Population Growth ▪ The Clean Nursery: Might As Well Start Right ▪ Midwives Help People Out ▪ Childbirth Isn't for Sissies ▪ Postpartum Care: You Deserve a Break ▪ Breastfeeding: They Drink Until They Pass Out ▪ Pump It Up ▪ Bottle-Feeding: Because We Can't All Be Milk Machines ▪ Vaccines: A Sticking Point ▪ Diapers: Be the Change ▪ Attachment Parenting in the Real World ▪ Talking to Kids About Saving the Planet ▪ *Who Knew?* Penguins and Polar Bears ▪ The Vegetarian Child: Trade You a Tofu Dog for Your Lentil Loaf ▪ Cooking with Kids: Because They Won't Eat It When I Make It ▪ Play Equipment: Swing on This ▪ Skip the Petting Zoo . . . Unless You're a Fan of E. Coli ▪ Birthdays: Towing the Party Line ▪ Waste-Free/Junk-Free Lunches ▪ School Buildings: Learning Where You Can Breathe

ten : clean rites of passage   173

Cradle to Grave ▪ Off to College: Sometimes We Have to School the Administration ▪ Setting Up Your First Green Apartment ▪ *Who Knew?* Wildfires ▪ Clean Dating: Why Diamonds and Roses Aren't Always So Romantic ▪ Weddings and Commitment Ceremonies (No, She Won't Wear It Again) ▪ Divorce and Child Custody: Staying Green Through the Transitions ▪ *Who Knew?* Oceans ▪ Green as You Gray: Eco-Conscious Retirement Communities ▪ Let's Finish What We Started: Green Burial and Cremation

Unless someone like you cares a whole awful lot, nothing is going to get better. It's not.

—Dr. Seuss, *The Lorax*

introduction

When a friend gave me a 300-page book on the ecological implications of laundry, I knew the environmental movement was doomed.

As a working mom, I hardly have time to *do* laundry—let alone read an epic tale about it. I wanted to call that laundry author and demand a synopsis: "Give me the two-minute version on dirty clothes."

As individuals and as parents, we have so many everyday issues to be concerned about—food, clothes, cosmetics, cleaning products, what needs to be organic, and what doesn't matter so much. It's easy to get overwhelmed. But then it occurred me. There are Cliffs Notes on the complete works of Shakespeare and Victorian novels reduced to Twitter feeds, so why not quick interviews with health and environmental super-brains to illuminate all we need to know in a few pointed moments?

I came to the environmental movement young, with big ideas about "saving the planet." My eighth-grade science teacher, Ms. Davis, taught us about the environmental impact of putting meat on our tables, so I became a vegetarian. It was only later that I found out what they put in American meat (no, you don't want to know, but I'll tell you anyway on page 10).

I signed on with the Ban the Box campaign in the eighties. Surely a few of you remember those giant cardboard boxes CDs used to come in—the "long boxes," impossible to open from any corner?

But here's the deal: Those boxes don't exist anymore. Activism works. We consumers saw all that absurdly wasteful packaging and said, "No. We want our Cyndi Lauper without killing a whole forest, thank you very much." It was an amazing feeling: We the people had the power to change the world, one long CD box at a time.

I was so excited I decided I'd really ramp up my activism. Yes, I'd join PETA in a protest against animal testing by a big cosmetics company! The spokesperson for the company was a famous actress, and we were protesting outside her house—with a man in a bloody bunny suit. As you can imagine, this didn't go over so well.

Thanks for forgiving me, Cybill.

I settled into less radical tactics and got involved with other causes, ones that were a little more low key. I became a vegan and committed to buying organic whenever possible. I fought against global warming and tried to save the seals.

But as years rolled by and my life got busier, sometimes it just seemed like there were so many damned catastrophes. My dreams of saving the planet ended up on the back burner.

Then I became a parent, and I started thinking about the future in a new way. What kind of world would my children grow up in? What about my children's children? When I found out that the flame retardants in my son's mattress, pillows, and pajamas were actually toxic to the brain, I went into a mild panic.

My son's bed was literally killing him?

How had I gotten too busy to save the planet? And if I couldn't save the planet, surely I could save my own kids. The same toxic choices that are killing the planet are making our homes unlivable and our bodies sick. I had to do something, but what?

Enter that 300-page book on laundry. What was I going to do with *that?* Use it as a doorstop? Donate it to the library?

The laundry book wasn't going to save the world, so I picked up another title, one that promised *simple* steps to a greener life. I opened to a random page and read that I could make a real impact buying candy without wrappers.

Okay. Pause. Take that in.

Candy without wrappers?

Sure, wrappers can be hard to recycle. But there wasn't a Ye Olde Candy Shoppe in my neighborhood. At least I could recycle the book.

. . .

I'm committed to being green, so if I'm overwhelmed, who isn't?

I think about my mom, about the people I work with—they're not happy about climate change or lead paint or obesity rates, but they're not excited about the environmental movement, either.

"Yes, Sara, I *am* recycling this," they promise, sort of defensively when I say "Good morning." As if the mere sight of me represents some kind of nag factor.

I don't want to nag anyone, believe me. And I know that if we don't make this easy, we'll never get back to the feeling and reality that we the people have the power to change the world.

Eighty percent of the world's forests are already gone.

Air pollution has gotten so bad that those of us who live in major cities live on average two to three years less than people who live in cities with cleaner air.

Thinking you'll just stay inside? Because of common household cleaners, pollutant levels in the average American home are actually two to five times higher than they are outdoors.

But this book isn't some Chicken Little routine about the current mess we're in. It's about starting the cleanup. It's clearly not about being perfect, either—there have been weeks when I didn't recycle—but then, I *did* use a hippie cleaning product to clean my kitchen. It's about doing what we can when we can and empowering ourselves to make a difference.

Having said that, I understand that the temptation to throw up our hands and say "Forget it" will always be there. I feel the allure of a shimmering new SUV just like the next mom. "I'm so sick of this small hybrid," I muttered under my breath as I climbed into my six-year-old Prius the other morning.

My almost-eight-year-old son was quiet in the backseat. I assumed he hadn't heard me. But at the first stop sign he made his appeal: "Mama, *please* don't get an SUV," he begged. "I can *see* the exhaust in the air."

I looked into the rearview mirror, into my son's wide eyes, and it hit me in the heart. *What have we done?* In the name of bigger and better, have we created a world in which our kids can hardly breathe?

Then it occurred to me: If there's anything we Americans like more than bigger and better, it's quicker, easier, and more entertaining. What if we had a new plan, one that's practical and informative, that draws the best real-world examples from top experts, everyday moms, and my own personal experience as someone who has struggled with these issues all her life? What if *doing something* could take less effort than the energy it takes to stay in denial? Well, we might actually have a solution on our hands.

Right then and there I decided to tackle that. If I could give people the bare essentials on each issue, the easiest ways to change our future—and our kids' futures— then our bodies and this planet might actually have a fighting chance.

If I couldn't get through the laundry book, I don't expect you to read *The Imperfect Environmentalist* from cover to cover. Even you, Mom. And I don't expect you to remember every fact in these pages. Instead, use this book as a resource when you want to paint your living room without poisoning your cat. Feel free to flip through the book to look up any subject you need to deal with—or just flip through when you feel like it. Pick a page and clean up your life one chapter, one topic, one page at a time. I would love for *The Imperfect Environmentalist* to be a conversation-starter, a source of inspiration, and a simple approach to help us clean out the pollutants, the poisons, and even the negativity in our lives. Let it be your head-to-toe, cradle-to-grave, tree-house-to-the-whole-crazy-planet, quick and dirty guide—the Cliffs Notes to illuminate all we need to know in a few pointed moments. One page per topic—even laundry.

Sara Gilbert
Los Angeles, 2012

the icons

air quality

animal rights

aquatic life

carbon footprint/greenhouse emissions

deforestation

ground pollution

happiness

health

local economy/community sustainability

money saving

parenting

renewable energy/energy saving

social justice/human rights

water quality

the **imperfect**
environmentalist

clean eating and drinking

Why don't we pay more attention to who our farmers are? We would never be as careless choosing an auto mechanic or babysitter as we are about who grows our food.

—Michael Pollan, journalist and environmentalist,
The Botany of Desire

We Are What We Eat

What we eat affects who we are, not just physically but emotionally and spiritually. For me, clean eating and drinking is about starting from the inside out. It's the environmental issue closest to my heart and the place to begin. I became a vegetarian because I'm strict and weird and like to control things. I was thirteen, so I felt good all the time anyway. Years later, when I became a vegan, I really felt it in my body. I immediately had more energy and watched an extra layer of fat just melt off. When I began having some stomach problems, I started seeing food as medicine and became macrobiotic—my stomach was suddenly fine, my skin got clearer, my energy became even throughout the day, and my overall mood improved. This isn't to say I'm not moody, but . . . not *as* moody. And to me that's worth eating all the berries and twigs I can find.

Even the Water's Not Safe?

Cut to the Chase, Hippie: What's the Least I Need to Know?

Safe drinking water isn't just something to worry about on your tropical vacation. U.S. tap water is ridden with arsenic, lead, and pharmaceutical drugs. In short: Get a filter.

Intriguing . . . I Can Handle a Little More

Tap water has been protected under the Safe Drinking Water Act since 1974, but experts warn that the water in our pipes increasingly doesn't meet health-safety guidelines. My tap water contains a horrifying thirty pollutants—including illegal levels of by-products from disinfectants, and dangerous amounts of arsenic and chloroform. How do I know? I went to the Environmental Working Group's website and typed my zip code into the "What's in your water?" tool. You can, too. Here's their website: http://www.ewg.org/tap-water/home.

So grab the bottled water off the grocery-store shelf? Not so fast. Is that glass or plastic bottle going to wind up in a landfill? And just because it's bottled doesn't mean it's chemical-free. If the water in your area is suspicious, stay hydrated by investing in a filter for your tap. Currently the two most popular types of filters are reverse osmosis and ionizing or alkaline filters. There are arguments in favor of both sides, but basically proponents of reverse osmosis filters claim that they get the water more pure by actually removing most of the chemicals. People who like ionizing filters say that because of the purification process, reverse osmosis water is overly stripped of minerals and therefore "dead" and acidic. Since the water doesn't have enough minerals, our bodies' own are leached as we process it.

Fans of ionized water also say that it is more hydrating and boast that it is alkaline, which is supposed to be a good thing, I guess. So far there are no real comprehensive long-term studies on which water is better, so go with your gut.

You can go to a hardware store or pick up a unit online where there are stores that specialize in water filters.

I chose a company called A Divine H$_2$O since they offer a system that combines reverse osmosis with an alkaline system that remineralizes the water, because I figure more is more. Plus they set positive intentions on the tank in the back of their store, so I mean, how can I not buy happy water? Once your tap water is filtered, take it with you in a reusable stainless steel or glass container.

If you do need to rely on bottled water, choose a brand in a glass bottle. There's something to be said for spring water since it's more "natural" than all of the filtered stuff. Just try to buy as local as possible. I mean, do you really need your water to come from Fiji?

I Need Some Facts to Bore My Friends With

Why no plastic? Well, most water bottles are made with plastics called "polycarbonates," which leak low levels of BPA into everything the plastic touches, including cool water. High BPA levels are associated with heart disease, diabetes, breast cancer, and abnormally high levels of certain liver enzymes. It's one of many chemicals classified as an endocrine disruptor, which is why polycarbonates are banned for use in baby bottles in Europe and Canada. Good news: As I write this, BPAs have been banned in the United States, too. Because, hey, if BPAs aren't good for Canadian babies, I'm going to bet they're not that great for the rest of us.

Organic Matters: You Are What You Eat

Cut to the Chase, Hippie: What's the Least I Need to Know?

Anything you're going to eat without peeling should be organic, but the inside of organic produce is better, too—with up to 40 percent higher levels of nutrients like vitamin C, zinc, and iron.

Intriguing . . . I Can Handle a Little More

My grandma used to say, "An apple a day keeps the doctor away," but maybe not when that apple's been treated with methylcyclopropene, a chemical gas that prevents fruit from ripening and makes that conventionally grown Fuji look fresh even if it's a full year old. Yes. A year old. The only way to be sure your produce is fresh and pesticide-free is to buy 100 percent certified organic food, which has to comply with regulations that severely limit the use of additives and fortifiers, and requires it be grown in a way that maintains the integrity of the soil and is in harmony with the larger ecosystem.

I Need Some Facts to Bore My Friends With

From personal health to global sustainability, organic food is truly at the heart of a sustainable lifestyle.

Every time you cook a meal with organic food, you're not only doing your body a favor but the earth, too. Organic farming reduces pollutants in groundwater and creates richer soil that aids plant growth while reducing erosion. Now, *that* is a really boring fact—don't pull it out on a first date.

I'm Donald Trump

Buy all organic all the time and eat only at certified organic restaurants.

Okay, I've Got My Own Place, but I've Also Got Credit Card Debt

If you're going to peel it, go ahead and buy the conventional produce, but when it comes to berries, make an organic vow—nonorganic blueberries have been found to be laced with as many as fifty-two different pesticides, and off-season strawberries are often shipped from countries with no pesticide regulations.

I'm Sleeping on My Friend's Couch and Eating Ramen Noodles

Wash nonorganic fruits and vegetables in a colander in the sink. Use a little baking soda and a nail brush and dry them off before eating. This will remove most of the pesticides and wax, making your produce safer to eat.

Eating Local: Because You Wouldn't Drive All Day to Pick Up a Tomato

Cut to the Chase, Hippie: What's the Least I Need to Know?

When eating out, ask your server what's local. Order it.

Intriguing . . . I Can Handle a Little More

Buying and eating locally grown food is good for the local economy because it keeps our dollars circulating close to home. It's environmentally friendly, because the food doesn't have to be transported long distances, requiring both fuel and preservatives. And local fruits and vegetables are better for our health—when food is picked ripe it tastes better and maintains its nutritional value.

I take my kids to the farmer's market—knowing I'm buying local and fresh makes the humiliation of the little choo-choo train ride worthwhile.

If you don't have a year-round farmer's market near you, ask your neighborhood grocer to offer more local options, or join a CSA—that's community supported agriculture, a popular way for consumers to buy seasonal food directly from area farmers. Go to http://www.localharvest.org for the low-down on where to find the best food grown closest to where you live.

I Need Some Facts to Bore My Friends With

According to a study by the New Economics Foundation in London, every dollar spent locally generates twice as much income for the local economy. When businesses are owned remotely, money leaves the community with each and every transaction.

Produce that's purchased in a supermarket has often been in transit or cold-stored for days or weeks. In comparison, produce at your local farmer's market has often been picked within the last twenty-four hours.

And it turns out that eating local is even more important for air quality and pollution than eating organic. A 2005 *Food Policy Journal* study found that the miles organic food often travels to our plate creates environmental damage that outweighs the benefit of buying organic. So go to that farmer's market, suffer the choo-choo train, and buy local *and* organic.

I'm-a-Better-Mom-Than-You Bonus

If you join a CSA, you can visit the farm once or twice each season. With that personal connection, parents find that kids favor food from "their" farm—even vegetables they've never been known to eat before.

who knew? climate change

Scientists are logging serious data showing significant geological and climatic change is happening over the course of *tens* of years, not tens of thousands. Lots of factors cause the earth's temperature to fluctuate, but when it comes to the issue of greenhouse gas concentrations, it's on us. It blows my mind that anyone still doubts that global warming is real. If you need proof, look at sea ice loss, vegetation growth changes, rainfall or its lack—and, for those who love hard evidence, ice-core analyses, which indicate the effect of rising sea temperatures through CO_2 variations. Think all this won't have much impact on anyone? An estimated 100 million people will be flooded by the end of the century. Not worried because it's a problem just for the future? According to the World Health Organization, 150,000 people already die each year as a direct result of climate change.

Eating Out: From Greasy Spoons to Fine Green Dining

Cut to the Chase, Hippie: What's the Least I Need to Know?

Stuck at an earth-hating restaurant? Even if there aren't specific local or organic options on a menu, you can reduce your environmental impact by ordering things lower on the food chain. For example, chicken has a lower environmental impact than beef, while sustainable seafood has a lower impact than chicken. Vegetarian and vegan items have the lowest impact of all.

Intriguing . . . I Can Handle a Little More

There are more certified organic restaurants than ever, and many feature seasonal and locally grown produce and hormone-free meat. Plan ahead—before you're faint from hunger, check out the Green Restaurant Association's online "find a restaurant" search engine for your local options: http://dinegreen.com/customers/default.asp.

On a good day, you could find yourself in a restaurant that serves all natural options; otherwise, staying green and dining out is a matter of strategy. For businesses, there is a big difference between incorporating organic items into the menu and actual certification, so many restaurants serve food that's somewhere between 100 percent organic and deep-fried with MSG. Many menus will offer just a few organic or locally grown dishes; these restaurateurs are proud of their organic offerings and will most likely highlight them on the list. If you don't see anything organic, ask for more information from your server. Your inquiry might result in just one green addition to your meal, but even the inquiry encourages restaurants to keep or add more natural choices. I personally ask so many questions at restaurants, I can only imagine what organic items have been put on my plate.

I Need Some Facts to Bore My Friends With

American restaurants throw away an estimated six thousand tons of food every day. All that food rotting in landfills contributes to global warming—see, when it decomposes, it releases methane, a greenhouse gas twenty-one times more damaging than carbon dioxide. So no matter what you're eating or where you're eating it, be conscious about how hungry you really are. And if you don't mind people looking at you like you're a total freak (yes, I've done it), bring your own to-go container for leftovers. Even if only your dog or chicken is going to eat your old pizza, it's better than letting it blow a hole in the ozone.

Where Does Your Meat Come From? Do You Even Want to Know?

Cut to the Chase, Hippie: What's the Least I Need to Know?

Deforestation, animal cruelty, dangerous work conditions in packing plants—the meat industry is a green nightmare at virtually every level. If you eat meat, buy it from smaller, local, certified organic sources—it won't contain added chemicals, antibiotics, or growth hormones.

Intriguing . . . I Can Handle a Little More

About 200,000 acres of rain forest are being destroyed every twenty-four hours, mostly for cattle ranching. So while cows munch their way across the Amazon, only to be slaughtered and turned into greasy hamburgers, nearly half of the world's species of plants, animals, and microorganisms are threatened due to deforestation. Speaking of hamburgers, because of the way large slaughterhouses operate, one study found that any single four-ounce patty was actually made up of anywhere from 55 to 1,082 different cows. Yes, you read that right. If that's not enough, the cattle industry is also responsible for 18 percent of the world's greenhouse emissions.

The global environmental picture aside, standard meat industry practices are cruel to both workers and animals. Workers in the meat-packing industry are paid little for dangerous jobs in close-quartered spaces littered with animal remains. In 2005, Human Rights Watch released a report called "Blood, Sweat, and Fear: Workers' Rights in U.S. Meat and Poultry Plants," which exposed basic human rights violations. It turns out that rates of injury, extreme temperatures, and denial of bathroom use are alarmingly commonplace.

As for the livestock, being raised for food means a life of torture. The factory farming industry strives to maximize profits by cramming animals into tiny, filthy spaces, drugging them to fatten them faster and keep them alive in deadly living conditions, and genetically altering them to grow bigger or to produce more milk or eggs.

When local organic meat isn't available, you might opt for a local Muslim butcher. Halal meat is raised and slaughtered in somewhat more humane conditions. Praying five times a day optional.

I Need Some Facts to Bore My Friends With

Of the thirty-two pounds of feed your average cow consumes in a day, 75 percent of that is corn. But corn isn't good for the land, and it isn't good for cows. First of all, corn is grown as a monoculture, meaning that the land is used just for corn and isn't rotated, which depletes soil nutrients, contributes to erosion, and ends up requiring more pesticides and fertilizer that, in turn, have been linked to oceanic "dead zones" and endocrine disruptions in animals, such as turning male frogs into hermaphrodites. As for the cows, their stomachs weren't made to digest corn. With the masses of corn being devoured, cows develop acidosis—or really terrible cow heartburn. "Acidotic animals go off their feed, pant and salivate excessively, paw and scratch their bellies, and eat dirt," says author Michael Pollan. "The condition can lead to diarrhea, ulcers, bloat, rumenitis, liver disease, and a general weakening of the immune system." Sound appetizing?

The Drive-thru: Going Nowhere Fast

Cut to the Chase, Hippie: What's the Least I Need to Know?

National fast-food joints sell a hydrogenated fat and high-fructose corn syrup dining experience—it's cheap in the short term, but it's deadly. If there's a healthier way to eat at a fast-food restaurant, it's probably to drink water and order a salad without dressing. Actually, hold the water, since that's in a plastic bottle; and might as well hold the salad, too, since that's probably modified and certainly full of pesticides.

Intriguing . . . I Can Handle a Little More

It may seem like the fast-food industry is in a world of hurt: After decades of propaganda, the truth is finally coming out that fast food kills. Studies show that its consumption increases Americans' risk of type 2 diabetes and obesity—which leads to cancer, heart disease, and other major health problems. Currently, about one-third of American children and almost two-thirds of American adults are overweight or obese. Shocking, right? But those national chains aren't suffering financially. Americans now spend more money on fast food than on higher education, personal computers, computer software, or new cars. And only a tiny fraction of that money supports our local economies.

Healthy food is packed with important nutrients: vitamins, minerals, antioxidants, all in low-calorie packages. But a burger and large fries at a fast-food chain can easily top thirteen hundred calories—and that's not even counting a sugary soda—providing hardly any vital nutrients. The burger alone supplies about 60 percent of your daily fat needs and 86 percent of your sodium. Even the wrappers at most fast-food restaurants are scary: that grease-proof paper breaks down into PFCAs: carcinogens that build up and remain in the body. So unless your idea of a good time is to have permanent hamburger wrappers lining your guts, pack your own lunch.

I Need Some Facts to Bore My Friends With

California's nearly three million high school freshmen are more likely to be obese if there's a fast-food restaurant within a block of their school, according to a new study by UC Berkeley economists, who calculated that these students eat thirty to a hundred more calories per school day.

Farmed Fish, Dyed Fish, Red Fish, Blue Fish

Cut to the Chase, Hippie: What's the Least I Need to Know?

Avoid shark, swordfish, king mackerel, and tile-fish because they contain the highest levels of mercury. Unless of course you're a fan of mercury.

Intriguing . . . I Can Handle a Little More

Most people know not to break open a thermometer and suck out the mercury, but we eat fish poisoned with the same heavy metal every day. To find out more about which fish to avoid and where to buy certified "fish to eat," check the Marine Stewardship Council's website: http://www.msc.org.

Fish and shellfish are low in saturated fat, and contain high-quality protein and omega-3 fatty acids—both of which decrease our chance of heart attack, make babies smarter, and ward off dementia and stroke as we get older—but because of mining and coal-burning practices, nearly all fish and shellfish contain traces of mercury. Too much mercury can affect our immune systems, alter our genetic and enzyme systems, damage our nervous systems, and lead to crazy diseases I can't even pronounce and you definitely don't want.

I Need Some Facts to Bore My Friends With

Thanks to overfishing and pollution, most species of tuna, swordfish, cod, salmon, and halibut are endangered, with noticeable population declines in rivers and oceans since the 1980s. With the threat of extinction, some conservationists turned their hopes to fish farming—raising fish in net pens in bays and channels—but farming has only made matters worse for many wild fish species.

It turns out that predator fish are high on the food chain, and are difficult to maintain in captivity because they require so many smaller wild fish to feed on. To promote breeding in captivity, fish farmers use hormones as well as a host of chemicals for other purposes, like antibiotics, pesticides, and disinfectants. Even fish farms designed to be "sustainable" threaten wild populations, because escaped cultivated fish change the gene pool of their wild cousins.

Still, there's some hope for sushi. In a few areas the government has managed the environmental practices well enough that populations of aquatic species are more protected. Freshwater salmon habitats in Alaska are still relatively pristine, because there's less river damming, deforestation and related runoff, and urban development compared to California and the Pacific Northwest.

Love Animals, Don't Eat Them

Cut to the Chase, Hippie: What's the Least I Need to Know?

There's plenty of evidence that vegetarians are healthier. Cancer, coronary artery disease, diabetes, obesity, gallstones, and kidney stones are all much less common in vegetarians. Plus we're just better. Ready to give up on meat and fish? Focus on what you can eat rather than what you can't eat. French fries or fettuccine Alfredo might not be healthy, but they *are* vegetarian. We'll worry about substituting soy milk for heavy cream in the vegan section.

Intriguing . . . I Can Handle a Little More

A plant-based diet is a healthy choice for virtually everyone, but as you make the transition, pay attention to your nutritional needs. The Mayo Clinic has developed a food pyramid for vegetarians: http://www.mayoclinic.com/health/vegetarian-diet/HQ01596/, and PETA offers a free vegetarian "starter kit": http://www.peta.org/living/vegetarian-living/free-vegetarian-starter-kit.aspx. In both, there are many variations on how to stay healthy with no meat.

You don't need to be an expert vegetarian cook to know the right foods to eat. Just do a little research on the basics. For example, it's important—especially for young women—to get plenty of iron. The type of iron found in beans, lentils, enriched grains, spinach, raisins, and fermented soy is enough, and eating citrus will help your body absorb it.

I Need Some Facts to Bore My Friends With

If we were all vegetarians, that would mean way less livestock on earth—and a healthier environment. Methane emission from livestock accounts for some 14 to 30 percent of methane gas emissions worldwide. Some experts estimate that one cow burping and farting all day creates as much pollution as your average car.

Millionaire-in-the-Making Bonus

The cheapest foods are plant-based. Studies show that the average vegetarian saves 20 percent at the checkout. Vegans save even more.

who knew? the return of the salmon

Salmon populations have dwindled down to 10 percent of original numbers. Such decreases signify that the quality, temperature, or cleanliness of the freshwater and river systems is compromised. Fish farms don't help—those farmed populations get back into the wild populations and make things even worse. If you must eat salmon, look for wild, troll-caught chinook, coho, sockeye, or other Alaskan salmon. "Fresh Atlantic Salmon" is another name for farmed fish, so don't be fooled into buying it.

Hug Me, I'm Vegan

Cut to the Chase, Hippie: What's the Least I Need to Know?

The science is with me. The China Study, a huge long-term project in rural China, found that a diet free of dairy products dramatically reduces our risk of cancer, diabetes, and heart disease. Start with selecting just one day a week—vegan Fridays, say. Doesn't that sound festive?

Intriguing . . . I Can Handle a Little More

I'd been a vegetarian for years when I learned two absolutely disgusting facts.

First, I found out that cheese isn't really vegetarian. A crucial ingredient in most commercial cheeses is a coagulating enzyme that actually comes from the stomach lining of baby calves. Check the ingredients: Rennet, rennin, and "enzymes" mostly come from veal or pork. Next, I learned that nursing moms who drink cow's milk may find their own breast milk coagulates, even plugging up in the milk duct.

That did it for me. Dairy products sounded about as appetizing as gelatinous pig guts.

Of course, any dietary change requires some planning and adjustment, but finding vegan alternatives to all our favorites is as easy as skipping down the grocery store aisle. Vegan pizza, vegan ice cream, and almond- or coconut-based milk alternatives are widely available, and many actually taste better than the dairy version—or at least that's what I'm telling myself. Daiya makes an awesome meltable cheese, and Coconut Bliss ice creams are vegan, soy-free, and agave sweetened.

I Need Some Facts to Bore My Friends With

New vegans often worry about protein, but plant-based proteins aren't hard to come by: Beans, tempeh, quinoa, nuts, and the endless grocery shelves of those strange pink "meat alternatives" are all high in protein. But there's no need to go crazy. A recent study found that all Americans, vegans included, get *too much* protein. Many of us actually have a harder time getting enough vitamin B12—if your vegan lifestyle leaves you feeling sluggish, try adding a little seaweed to your diet.

Thanks to dairy-industry advertising and lobbying, most Americans think milk "does a body good," but cow milk is designed for baby . . . *cows*, and most humans on the planet don't drink the stuff. The National Research Council reports that cow's milk contains all the pesticides and hormones cows ingest with the alfalfa they eat. *Yum, pesticides.*

But enough depressing news. Going vegan won't only lighten your carbon footprint and make you feel better, it'll make you *look* better, too. Because, let's face it, eating pesticides, growth hormones, and stomach lining just aren't that conducive to beauty.

The Raw Deal: Why Bother Cooking?

Cut to the Chase, Hippie: What's the Least I Need to Know?

Raw foodists believe that heating food above 118 degrees can destroy natural enzymes, and when our bodies have to produce those same enzymes for digestion, that's diverted energy that would otherwise be used for healing and cleansing. Start by adding one raw element to every meal—and, no, that lemon wedge in your diet soda doesn't count.

Intriguing . . . I Can Handle a Little More

A raw food diet doesn't just mean chewing on greens and seeds. Raw chefs use ingredients rich in nutrients and antioxidants to create soups, salads, lasagna, veggie fajitas, pesto, pizzas, and brownies, among other things.

Sprouting is key to preparing many food items that are traditionally thought to need cooking in order to make them digestible. A raw foodist will simply soak the seeds, nuts, grains, and legumes, causing them to germinate and sprout. The sprouts are packed with bioavailable vitamins, minerals, amino acids, and proteins.

Many raw cookbooks suggest having a blender, food processor, juicer, and dehydrator for making raw recipes, but you can get started without any fancy equipment at all. Keep it quick and simple by choosing not to imitate cooked food; instead, make dishes like Thai salad, smoothies, and guacamole.

I Need Some Facts to Bore My Friends With

Studies show that a raw diet lowers obesity and cholesterol, and even reduces symptoms of rheumatoid arthritis and hypertension. Whether a raw diet is a temporary choice or a long-term way of life, make sure you're getting enough of vitamins D and B12 and you'll likely notice within a few days that you have more energy.

I'm Donald Trump

Invest in a top-of-the-line dehydrator and hire Ani Phyo as your private raw chef.

Okay, I've Got My Own Place, but I've Also Got Credit Card Debt

Pick up a few beginner's raw cookbooks at your local bookstore.

I'm Sleeping on My Friend's Couch and Eating Ramen Noodles

Check out the blog "Raw on $10 a Day (or less)" for cheap raw food menus: http://rawon10.blog spot.com.

Macrobiotics: Extreme Balance

Cut to the Chase, Hippie: What's the Least I Need to Know?

Macrobiotics is a philosophy and way of life that requires study and practice—even though I am one, I don't understand it all—but the dietary basics are simple: Eat a balanced diet of in-season, locally grown vegetables, grains, beans, and fish protein based on your condition. And chew your food really well.

Intriguing . . . I Can Handle a Little More

In Latin, "macro" means long and "bio" means life—so the goal here is to eat for a long and healthy life. Followers of the traditional macrobiotic approach believe that food has a profound effect on our health, well-being, and general happiness. I went to a macrobiotic counselor when I became macrobiotic. Following her principles, I felt an immediate effect on my energy level and overall health. But I drew the line when I had a fever and she told me to put a piece of tofu on my head.

People who follow a macrobiotic lifestyle might have trouble ordering from most restaurants, so we usually put together our meals at home. There are various principles involved, from sitting down during meals and doing nothing else but eating to taking a stroll at the end of a meal and being more active in general. As I'm sure you've guessed, the quality of food becomes extra important for a macrobiotic, so organic produce and health food stores are your friends.

I Need Some Facts to Bore My Friends With

Macrobiotics is strictly disciplined, down to the order in which you eat your food: At every meal, heartier foods come first, followed by vegetables and salads. The system breaks foods down into whether they are "yin" or "yang," and a macrobiotic counselor can diagnose if an individual is too yin, too yang, or balanced. Being artistic and creative generally indicates that you're more yin, but my counselor was quick to point out that in spite of this I'm still more yang because I'm short. We also strive to balance our blood. We don't want it to be too acidic or too alkaline. The American diet is very acidic—caffeine, sugar, and alcohol all contribute to that. So many Americans would be advised to lean toward more alkaline foods such as grains, beans, and vegetables.

As is common with other types of dietary guidelines, people who are macrobiotic try not to eat anything within three hours of bedtime. It's not just about food, either—macrobiotics ideally means going to sleep by midnight, waking early, and even extends to outer stimuli, such as your home. A natural living space, full of plants, natural fibers, and all-around clean aesthetics makes for a happy macrobiotic. Further, macrobiotic practitioners try to constantly challenge themselves to grow, learn, change, and avoid overall stasis. This applies to diet—don't eat the same foods every day—and to life—stay curious; for instance, don't mindlessly take the same walk day in and day out. Speaking of walking, my counselor mandated that I walk barefoot in the grass outside but not at home when it's cold—don't ask why, because honestly I don't know, but I'm going with it.

Soy Happy

Cut to the Chase, Hippie: What's the Least I Need to Know?

Instead of using tofu for three meals a day or buying prepackaged fake meat, try miso and tempeh. Both are delicious and ancient ways to eat soy, made from fermented beans—which means they're healthier. And what doesn't sound appetizing about fermenting a bean?

Intriguing . . . I Can Handle a Little More

Touted as a perfect miracle food by the FDA and vilified by many for causing health complications, including allergies, increased estrogen in men, and thyroid problems in women, the soybean has been grown for food in Asia since before written record.

Many of soy's health benefits have been linked to isoflavones—plant compounds that mimic estrogen. But animal studies suggest that eating large amounts of those estrogenic compounds might reduce fertility in women, trigger premature puberty, and disrupt development of fetuses and children.

The controversy stems from the fact that not all soy is created equal. Soybeans can be cultivated using organic techniques or mass produced in ways that are harmful to the environment and result in low-quality foods. Look for soy products that are specifically labeled organic or GMO-free—meaning the soybean hasn't been genetically modified.

I Need Some Facts to Bore My Friends With

Soy flour, concentrates, and isolates are sketchier food choices than organic GMO-free whole soy beans and fermented foods. The lower-quality soy is commonly used in imitation meat, providing the right texture for products with amusing names like Soysage, Not Dogs, Fakin' Bacon, Sham Ham, and TofuRella. The fact that soy isolates are made using much the same technology as plastic toy parts is less amusing, as are the potential negative health effects. Another culprit when it comes to soy is phytic acid, which binds healthy minerals together so that the body can't absorb them. That's where fermentation comes in—it neutralizes the phytic acid. The results include miso soup, a Japanese culinary staple; miso paste, which can be made into sauces and spreads for vegetables; and tempeh, which is similar to tofu in that it can be used in dishes that usually call for meat.

I'm Donald Trump

Tell your personal chef to substitute tempeh for the beef in your lo mein.

Okay, I've Got My Own Place, but I've Also Got Credit Card Debt

Buy GMO-free and organic soy products at your local grocery store or co-op and experiment with new vegetarian recipes.

I'm Sleeping on My Friend's Couch and Eating Ramen Noodles

Swap out the ramen for miso soup with organic brown rice.

Grocery Shopping: Beyond Paper or Plastic

Cut to the Chase, Hippie: What's the Least I Need to Know?

A plastic shopping bag can take anywhere from fifteen to a thousand years to decompose. Avoid disposable plastic and paper bags by bringing your own reusable cloth grocery bags to the market.

Intriguing . . . I Can Handle a Little More

There are plenty of ways to stay green-conscious at the grocery store. Start by walking or biking to the store to cut back on traffic congestion and pollution. If you do drive, carpool with a friend or two. Buy just what you need—and make a shopping list before you go so you don't have to make another trip if you forget something.

As you shop, think about wasteful packaging: Avoid individually packaged items and look for recycled packaging.

I Need Some Facts to Bore My Friends With

Organic food is often criticized for its cost, but small and natural grocery stores are rarely criticized for the way they treat their workers or the environment.

Consider some facts on big box stores before you blow off the little guy for charging too much: Superstores destroy animal habitat and increase traffic. Whenever these giants open their doors for business, nearby locally owned shops lose income and often go out of business completely, forcing everyone in a small town to work in one large corporately owned business.

This might not seem so bad if these stores treated their employees well, but workers are usually not unionized, and may even be discriminated against. Recently the Supreme Court heard a case in which a major company was allegedly denying promotion to workers based on gender. (Guess which store. Guess which gender.)

Larger stores usually get their products from farther away, too, meaning more packaging and fewer bulk items.

Listen, you may spend a little extra buying organic produce from an environmentally friendly grocery store, but maybe it's worth it. As I get older I notice a lot of my peers have started to need to take medicine for some ailment every day, and that's expensive. So, sorry for the nerdy expression, but maybe an ounce of prevention is actually worth a pound of cure.

GMOs: What Did They Just Do to That Plant?

Cut to the Chase, Hippie: What's the Least I Need to Know?

Genetically modified foods may increase deadly allergies in children, have been shown to inadvertently damage butterfly populations, and threaten to financially devastate traditional farmers around the world. So choose food that's specifically labeled GMO-free, even if it isn't pretty. And if you can't find that label, look for food that is labeled 100 percent organic, because under U.S. law anything with that label cannot contain GMOs.

Intriguing . . . I Can Handle a Little More

Most Americans eat genetically modified foods without even knowing it. As much as 80 percent of all packaged foods contain GMOs, which the labels don't have to tell you.

Genetically engineered plants may look better on the shelf because they are larger, often freakishly so, and have fewer visible flaws. Oranges can be engineered to be brighter in color, and be higher in sugar content, but choosing a designer orange instead of an organic one is like choosing a candy bar. It may be sweet, but it isn't worth the risk; scientists are studying the effects that engineered plants have on our bodies, and early findings don't sound good. Soy allergies skyrocketed by 50 percent in the United Kingdom soon after GM soy was introduced, hundreds of laborers in India reported allergic reactions from handling GM cotton, and lab rats that are fed GM food have a wide variety of liver problems.

If that's not enough, the patented GMO seeds from huge corporate farms blow over to neighboring small farms and the big companies sue the small farmers for growing their patented product even though it was carried over by wind. These ridiculous lawsuits force the smaller farms out of business because they can't keep up with the legal fees. Each time a natural farm is muscled out, we lose access to non-GMO food.

I Need Some Facts to Bore My Friends With

Environmental activists, religious organizations, public interest groups, professional associations, scientists, and government officials have all raised concerns about GM foods and criticized agribusiness for pursuing profit without concern for potential hazards. Still, the U.S. government has exercised little regulatory oversight. Monsanto, whose dangerous corn seeds are banned throughout Europe, currently dominates the GMO seed market right here in the good ol' U.S. of A.

Juice This

Cut to the Chase, Hippie: What's the Least I Need to Know?

Trendy but true: Juicing is a good way to get more fruits and vegetables into our bodies. Health professionals recommend two or three cups of vegetables a day, and throwing all that produce into a juicer can make your intake move faster. Cutting out the fiber, however, may not be the best idea when it comes to digestion and absorption. People at risk for diabetes should be particularly careful.

Intriguing . . . I Can Handle a Little More

Some people detox with juice for a day or two to rid their bodies of toxins and give their organs a break. Drinking fresh juice at the start of the day, eating raw fruit and vegetables for lunch, and a light evening meal is a gentle way to detox. An actual juice fast is stricter and usually means some weight loss.

Some popular ingredients for juicing are apples, carrots, beets, pineapples, raspberries, cranberries, celery, cucumbers, oranges, spinach, and other greens. Adding herbs, lemon, or ginger can spice things up.

I Need Some Facts to Bore My Friends With

Studies have shown that juicing fights cancer, boosts the immune system, slows down aging, and is a natural and easy way to lose weight. Just watch the sugar intake.

I'm-a-Better-Mom-Than-You Bonus

Take your kids to pick-your-own fruit and vegetable farms—and juice the harvest.

Or try making apple juice at home and adding some kale and spinach. If the choice is between juice and water, your kids might just start loving blended kale.

Beauty Bonus

People who drink fresh juice daily say they have better, brighter skin. I haven't noticed the glow myself, but, hey, if they say they look better . . .

Caffeinate Me: All About Coffee

Cut to the Chase, Hippie: What's the Least I Need to Know?

My coffee place opens at 7:30 and I'm there at 7:29 every day waiting for my soy latte. I love coffee as much as my barista hates me. Americans agree with me, and we use some 23 billion disposable cups every year, which come from clear-cut forests and end up in overflowing landfills. If you can't brew coffee at home, bring your own travel mug to the coffee shop.

Intriguing . . . I Can Handle a Little More

Once you've got your sustainable cup, ask your barista for organic coffee, so you're not poisoning the earth just to wake up. Coffee is a cash crop—after oil, it's the second most highly traded commodity in the world. In this multibillion-dollar industry, it's no surprise that coffee is mass produced, usually by agribusiness conglomerates that use chemicals to ward off common pests. While the roasting process may dilute or elimi- nate the harmful effects of these chemicals for consumers, coffee workers and their families are at high risk and the environment suffers.

Even better than organic coffee is "fair-trade" organic—meaning the coffee has been purchased directly from the growers at higher prices, provid- ing economic incentives for sustainable produc- ers and promoting healthier working conditions.

I Need Some Facts to Bore My Friends With

Because coffee is grown only in two places in the United States—Hawaii and Puerto Rico—most Americans can't "buy local" when it comes to cof- fee beans, but we can still be conscious of how our coffee is grown.

Save trees by opting for "shade-grown" coffee when you can—conventional coffee farming often means that trees are cut down, coffee is planted, and within a few years the soil is depleted and the farm has to move to a new location, meaning more deforestation and habitat devastation. "Shade-grown" coffee is, well, grown in the shade—usually on small farms, using sustainable techniques.

Tea Time

Cut to the Chase, Hippie: What's the Least I Need to Know?

Tea is less acidic and less caffeinated than coffee, so it's healthier, and because it's both stimulating and relaxing, it won't give you the jitters. Therefore, less fun.

Intriguing . . . I Can Handle a Little More

Green tea has long been thought to protect against some types of cancer, and there is now an increasing amount of scientific evidence showing that it can prevent Alzheimer's, cataracts, colon cancer, and even help women conceive. Researchers in Britain and Italy found in lab tests on rats that black tea protected against the effects of a known carcinogen. Drinking tea regularly can also improve heart health and help with weight loss. Of course, all of these health effects are dulled after three sugar cubes and a cup of milk, but we can't all be the Dalai Lama.

I Need Some Facts to Bore My Friends With

Loose-leaf, fair-trade, organic tea is the crème de la crème of tea. Black, oolong, green, pu-erh, or white, if the quality of the tea is high, you'll find you don't need to add anything to it. All of these varieties are made from the same plant, *Camellia sinensis,* and differ in appearance and taste only because of how they're treated after harvest. Black tea leaves are oxidized, or fermented, and white tea leaves are simply the dried buds of the tea plant. Because I go as granola as possible, I drink kukicha tea, which is actually made from the twigs of the *Camellia sinensis* plant. It's rich with vitamins and minerals and has the least amount of caffeine.

Herbal tea, on the other hand, is made from a variety of plants and may have medicinal purposes. Valerian, for example, helps with sleep, fennel helps with respiratory problems, cinnamon balances blood sugar, and ginger relieves morning sickness, migraines, and even arthritis. Seems worth trying before you pop a pill.

The Jolly Green Drinker: Beer, Wine, and Spirits

Cut to the Chase, Hippie: What's the Least I Need to Know?

By shipping their products far, U.S. alcohol manufacturers release the annual equivalent of 1.9 million households' greenhouse gas emissions—so buy regional beer, wine, and booze.

Intriguing . . . I Can Handle a Little More

One fine point to consider is that shipping cans produces 30 percent less greenhouse gas emissions than shipping glass—and kegs are even more efficient. Who knew that college frat parties could be so sustainable? If the kegger isn't your scene, BeerTown.org features a "brewpub locator" to help you find local small-scale beer producers.

In addition to shipping repercussions, it's sobering to consider what's in your alcohol and how it's produced. But there are plenty of companies that compost, and use organic ingredients and renewable energy—one company even plants a tree for every bottle of gin you buy. If that's not an incentive to drink, I don't know what is. Most organic beverages are what they claim to be, but in the wine industry many of the eco-friendly efforts are still self-monitored, so it can be tricky to be sure from labels which wines are really organic. Your safest bet is to go for the biodynamic wine that is certified by a third party. Biodynamic agricultural practices are even stricter than organics when it comes to sustainability. Only want that wine if you're sure it was racked under a new moon? Biodynamic it is.

I Need Some Facts to Bore My Friends With

Looking for the wino-of-the-year award? The greenest alcohol of all might be the stuff you make yourself. Home beer- and wine-making gives you the freedom to concoct a beverage tailored to your palate, and you can reuse glass bottles. This often requires very little in start-up costs, and there are many companies that sell supplies for homebrewers, including a number that offer organic ingredients.

If you end up making the stuff all day every day, I can introduce you to this guy, Bill W.

Chocolate Without the Child Labor, Please

Cut to the Chase, Hippie: What's the Least I Need to Know?

Chocolate's like coffee: addictive, delicious, and often harvested unfairly and loaded with pesticides. It doesn't have to be like that, though. Look for organic, fair-trade chocolate with a high percentage of cacao for a sweet way to absorb antioxidants.

Intriguing . . . I Can Handle a Little More

West Africa's Ivory Coast exports nearly half the world's cocoa, but it's made a bad name for itself with heavy reliance on child labor. Most supermarket chocolate either originated here or at the very least is tainted with harmful pesticides. Never fear: All health food stores and many supermarkets carry lovely chocolate bars made from high-quality cacao beans and harvested and produced with fair labor. There aren't many choices in terms of domestically produced chocolate—Hawaii is the only state with large-scale cocoa production—but many locations in Central and South America grow organic, fair-trade chocolate.

Even better than fair-trade, direct-trade growers are paid as much as four times their fair-trade counterparts. Both types of trade are labeled on the chocolate bar, so look closely before you buy. There's also eco-trade, which is similar to fair-trade, with strict regulations concerning working conditions, wages, and child labor. All three—direct-trade, eco-trade, and fair-trade—pay farmers reasonable prices for their goods. And some chocolate-production companies support slow harvest, which avoids clear-cutting rain forests and focuses on adequate pay for the individuals who yield the beans. Look for Rainforest Alliance–certified chocolate, which is shade-grown in old-growth rain forests.

For the most part, organic chocolate means it was fairly harvested and produced. Like almost everything else, if the food is organically grown, it's likely that its existence, from farm to table, is healthier for the planet—and healthier for you, too.

I Need Some Facts to Bore My Friends With

Contrary to popular belief, chocolate doesn't give you that natural high because of caffeine, but rather because of another naturally occurring stimulant, theobromine, which is a closely related chemical compound. The darker the chocolate, the higher the levels of theobromine. Unlike caffeine, theobromine is gentler, smoother, nonaddictive, and long lasting. Some say it's a mild antidepressant—anyone who's indulged in a piece (or two or three) on a stressful day would probably agree. Nothing like theobromine to take the edge off.

Sugar Isn't So Sweet

Cut to the Chase, Hippie: What's the Least I Need to Know?

We all know most cookies contain sugar, but sugar is also added to lots of prepared foods—like salad dressings, for instance. And sugars are best in extreme moderation, because the more added calories from sugar, the more risk of weight gain, Type 2 diabetes, and high triglyceride levels.

Intriguing . . . I Can Handle a Little More

As is the case with most harvestable delicacies, mainstream sugar production guzzles fossil fuel, water, and pesticides. Moreover, by the time it reaches you—whether in crystalline form or already added to brownies or pie—it's not going to do your body and mind any favors.

So while all sugar is best enjoyed as a bonus, not a staple, the good news is that there are some sugar companies that support eco-friendly growing practices, fair-trade harvests, and less processing. Look for fair-trade stickers for peace of mind; the best of these sugar-production companies employ the methods of seed variation, crop rotation, pesticide-free fertilization, zero-discharge wastewater management, and sustainability.

Beet sugar, from—you guessed it—sugar beets, can be found domestically grown (parts of the upper Midwest house sugar beet farms), which means the final product sits on a truck or a shelf for less time than its imported cousin. Cane sugar farms can often be found in the United States, usually in warm southern climates, such as Florida. Both beet and cane sugars are typically less processed and easier on the body, making them your best bet if you must ingest sugar at all. I prefer maple syrup or brown rice syrup. But I realize not everyone shares my views. When I offered to bring water or coconut water to a party at my son's school as an alternative to sugary drinks, another parent commented, "only if the kids are fasting or doing a cleanse." I decided birthdays were as good a day as any to cleanse, and as a bonus uphold my role as class buzzkill.

I Need Some Facts to Bore My Friends With

On the whole, the sugarcane industry uses much fewer fossil fuels than, say, the corn industry, so, if you must, always reach for some form of sugar (cane, beet, or even white) instead of high-fructose corn syrup. Also, after much debate about whether bodies process corn syrup in the same way as sugar, researchers at Princeton University found that rats who consumed corn syrup gained significantly more weight than rats who ingested the calorically equivalent amount of table sugar. While the results of the study have been questioned by some and the great corn syrup debate rages on, I say why not avoid the stuff? Even if it's not worse than sugar, good luck finding a credible source who will tell you it's *good* for you. Unfortunately, because the government subsidizes corn growers, it's cheaper to eat a candy bar than it is a bunch of broccoli—but that doesn't make it better. The same goes for ethanol: Rather than ethanol made from corn, sugarcane ethanol is fast becoming a better option as a renewable fuel. The Environmental Protection Agency calls ethanol made from sugarcane an Advanced Renewable Fuel, a designation given to biofuels that will encompass many billions of gallons of this country's fuel supply over the next ten years. Sugarcane ethanol gives off less carbon than gasoline, which reduces greenhouse emissions. It's also considered renewable, sustainable, and energy efficient.

Nontoxic Food Preparation: Would You Eat Your Cutting Board?

Cut to the Chase, Hippie: What's the Least I Need to Know?

While you're julienning your organic farmer's market carrots, your cutting board might be leaching toxic chemicals into that pristine orange flesh. The best kitchen products are BPA-free, dishwasher-safe, and recyclable.

Intriguing . . . I Can Handle a Little More

A quick search of "eco-friendly kitchen products" will yield far more items than you'll ever be able to use in one lifetime, from table linens to cookware and everything in between. One rule of thumb: Wood is good. As long as it's not coated with toxins. Wooden butcher blocks are longtime staples in eco-conscious and gourmet kitchens alike, and it's easy to see why: They are more durable than the plastic versions, and they don't give off harmful chemicals. The same can be said of other wooden kitchen utensils. This goes for glass, too, which is the better choice when handling meat, since wooden cutting boards can harbor bacteria. Bamboo, quickly becoming wood's cutting-edge, environmentally friendly twin, is also nontoxic and resists bacteria.

As to the rest of the kitchen and dining room, common sense is paramount. Glass, wood, and bamboo will always be safe options; plastic may or may not be, depending on whether it contains BPAs and other chemicals. Lovely recycled-glass serving bowls, stemware, and dinnerware exist, which is the kindest option for the earth and a good conversation starter at dinner parties if you're eating with really boring people. When it comes to placemats, while most fabric is relatively benign, look for linens made with fair-trade fabric and fruit or vegetable dyes for the most conscientious table setting possible.

I Need Some Facts to Bore My Friends With

The greenest of green home chefs know that meals begin at the supermarket, so shop with reusable bags—including reusable produce bags. Find them at Ecobags.com—where else? Even if you buy the cleanest kitchen supplies in the world, you could be negating some of that by using lots of plastics at the grocery store, which might contaminate your vegetables even while they sit in the fridge. At any rate, the widespread use of plastics on our planet is detrimental at the very least, so relying on paper bags and reusable produce containers is your best bet.

clean house

People will always try to stop you from doing the right thing if it is unconventional.

—Warren Buffett, business magnate and philanthropist

Your Own Personal Green House

I rent a totally green house built by green contractors. People often think they'll have to change so much to live green in their houses; but honestly, there's no inconvenience. I walk on reclaimed wood floors, the house sports a rain catchment system, and the shower is rigged with an energy-saving gizmo that keeps my waste—and my water-heating bill—to a minimum. It feels great to know I'm living in a clean environment, and my bills are much lower than they'd be in a conventional house.

The problem . . . I'm moving. The new house was recently redone so it had to follow certain new green building standards but it's certainly not LEED certified. While it has a cool roof and nontoxic flooring, there's low- but not no-VOC paint everywhere, the ceilings are covered with a thousand energy-guzzling recessed lights, and the inspector said something boring about the stratification of heat. Dilemma . . .

What do I do to make this existing imperfect environment better? I'm going to use a HEPA filter with a carbon VOC filter upgrade, then delete some of the lights, and weatherproof—starting with closing up the enormous gap some brilliant door installer left around the front door.

From Tree Houses to McMansions: How Much Space Do You Need?

Cut to the Chase, Hippie: What's the Least I Need to Know?

While 15,000 square feet might sound luxurious, most current home buyers actually want to downsize, according to the National Association of Home Builders. If you were feeling bad about not being able to afford more space, let this chapter be your silver lining.

Intriguing . . . I Can Handle a Little More

The bigger-is-better principle popped with the housing bubble and rising awareness about green building. Makes sense: Smaller spaces require fewer resources, which means saving money and saving the planet. If you're in the market for a new home, keep green options a priority rather than square footage. A house already outfitted with solar panels or water catchment devices will be more worth your while (and your money) in the long run than a larger spot that guzzles gas, oil, and electricity. This becomes even more true if you're building your own house—while it's not cheap to install solar panels, it'll save money over time, and the cost can be lessened by slicing off a few hundred square feet from the finished proj-

ect. You won't miss the second guest room or the annoying second guest who always seems to show up to fill it.

I Need Some Facts to Bore My Friends With

There are more than 7 billion of us on the planet. That's 4 billion more people than there were just a half century ago. Before you build your dream house, consider who you'll be evicting—any plot of undeveloped land has its own ecosystem, which becomes just a little more unbalanced each time a new house or development breaks ground.

who knew? habitat destruction

Those nature sounds you hear may be the deer, birds, and bears telling you to get out of *their* backyard. Whether it's the sprawl of expanding suburbs or the luxury log cabin in the woods, all that infrastructure is altering the natural forests, mountains, wetlands, and other spaces where wild things have lived and roamed for centuries. The U.S. National Forest Service cites habitat destruction as the leading cause of species endangerment, with studies finding up to 82 percent of federally listed species affected. Logging, water development, mining, grazing practices, and recreation are the main culprits.

Home Inspections: What to Ask

Cut to the Chase, Hippie: What's the Least I Need to Know?

Found the perfect new home? Or just want a checkup on your existing place? A certified American Society of Home Inspectors agent can tell you a lot about how green the place is—from finding leaks and cracks to sniffing out malfunctioning appliances. Better yet, contact BPI Building Performance Institute for a certified inspector who will performance-test your entire home.

Intriguing . . . I Can Handle a Little More

A home inspector who is also able to conduct an energy audit is your greenest bet. Home inspections are usually done just to ensure that a building is up to code and all appliances and aspects of construction are functioning properly. But for a little extra cash, all of the green features of a home can be confirmed and tested in an energy audit. If a seller claims that the kitchen cabinets are made from low VOC (volatile organic compounds) ma-

terials, the energy auditor can tell you if the claim is bogus. The auditor also finds information about heating and cooling, water and air leakage, insulation, water heaters, windows, and the like. He or she can tell you how effective the home design is in saving energy, which translates to saved cash. Find out which external forces could be compromising the efficiency of decks, doors, siding, windows, and other outdoor surfaces. While your inspector looks for leaks and cracks in the roof and drywall, have him check gutters and downspouts.

You can even ask an energy auditor to evaluate energy bills for the past year, and determine how the current appliances are doing and whether, say, your fridge should be replaced because it's using too much electricity.

I Need Some Facts to Bore My Friends With

Knowledge is power: Gas and carbon-monoxide leaks don't have to be dramatic—plenty of homes have slow leaks that aren't deadly but can cause cold-like symptoms or inertia. Come to think of it, maybe a slow carbon-monoxide leak is to blame when I'm inert.

Green Design and Construction

Cut to the Chase, Hippie: What's the Least I Need to Know?

According to the U.S. Department of Energy's Building America Program, a green design can reduce the energy consumption of a new home by 50 percent—with little or no impact on the cost of construction. So if it's time to break ground, don't source mahogany from Asia and support rainforest logging. Build the whole thing with local materials, and keep an eye on things like how many lights you really need.

Intriguing . . . I Can Handle a Little More

Every geographical area has its own ecosystem—so build your place with local resources and local conditions in mind. A desert home can be constructed with fire-resistant material and a roof that collects what little rainfall there is. In the forest, a deck might be built with a hole in it to accommodate a hundred-year-old tree, which will shade the house and keep it cool in the summertime. If builders remove grass to make room for a home's foundation, they can save it and plant it on the roof for extra insulation. If local, salvaged, or recycled wood is used for the frame, it can be used again if the house is ever renovated or demolished.

I Need Some Facts to Bore My Friends With

Many communities have stores that sell surplus house paint, plywood, tiles, and other building basics that are leftover from previous projects. Steering clear of vinyl, plastic, and other unrenewable resources might mean a slightly higher bill in the end, but it'll also mean you're not inhaling the outgassing toxic chemicals that many of these products produce when you move into your little sanctuary.

I'm Donald Trump

Hire one of the nation's premier green architects to create a zero-carbon-footprint mansion, and remember to keep the square footage to a minimum and the salvaged materials plentiful. Just because it's recycled doesn't mean it has to be made of beer cans—you can find marble from an old house or stained-glass windows that were replaced after someone else's renovation project. To keep that carbon footprint in check, ask your builders to use only materials that are obtained within a hundred-mile radius.

Okay, I've Got My Own Place, but I've Also Got Credit Card Debt

Consider recycling someone else's trash into your treasure by using reclaimed flooring. You can also keep your eye on energy waste by making sure you weatherproof all doors and windows well. This will save on that electric bill, and give you the extra cash you need to finally pay off those credit cards.

I'm Sleeping on My Friend's Couch and Eating Ramen Noodles

Help your friend build by hunting down salvaged supplies to use in construction. And if you're looking for a new career, check out the green-building classes at your local community college.

Hitting the Floor

Cut to the Chase, Hippie: What's the Least I Need to Know?

It might feel good between your toes, but that plush carpet could be invisibly outgassing heaps of toxic chemicals. If you are stuck on carpet, go for natural fibers that haven't been sprayed with a slew of toxins. Your best bets for flooring are bamboo, cork, concrete, or linoleum.

Intriguing . . . I Can Handle a Little More

A study published in the scientific journal *Neurotoxicology* linked vinyl—or PVC—flooring with autism. Vinyl emits chemicals known as phthalates, and though the research isn't conclusive, the findings are unsettling. PVC may also be a factor in medical problems ranging from asthma to sperm damage to certain cancers. And PVC isn't the only problem—carpet can be a veritable petri dish of dust, dirt, and allergens.

So let's look at some better options. Hardwood flooring is renewable if responsibly harvested, but stands of the trees might take up to fifty years to fully replenish after replanting. "Rapidly renewable" flooring, on the other hand, refers to materials that regenerate within about ten years. If you are going this route, you might try DuChateau flooring. I just discovered them since they were used in my new house. They qualify for LEED certification for being rapidly renewable and sustainably harvested, and because they have low emissions. Just be sure to steer clear of cut-rate, irresponsibly raised hardwood. Clear-cutting practices often destroy entire tree populations—forever.

Bamboo has become increasingly popular, and it's about as durable as oak. Bamboo planks resemble wood, but the stalks typically regenerate within three to five years. On the other side of the spectrum, linoleum might seem a little low-rent, but it's actually made of all-natural ingredients—linseed oil, wood, and cork. You can actually also make flooring out of recycled wine corks. Now you know what to do with that collection.

I Need Some Facts to Bore My Friends With

Look for lumber and building materials deemed worthy by the Forest Stewardship Council or similar organizations. Certification from the FSC requires that traditional hardwood and softwood flooring materials are sourced from operations that promote healthy ecosystems and local cultures and economies. And if you go with wood flooring try Rubio Monocoat for a plant-based nontoxic finish. Because if you're going through the trouble of avoiding the toxins in that carpet you didn't choose, the last thing you want to do is coat your floor in chemicals.

Raising the Roof

Cut to the Chase, Hippie: What's the Least I Need to Know?

Roofs don't have to be all tar and shingles. Go for a "green" roof, a "cool" roof, a roof that catches rain, or even a roof made from recycled tires.

Intriguing . . . I Can Handle a Little More

A green roof is quite literally green: Also known as an eco-roof, it's a conventional roof covered with vegetation, which in turn creates air-cleaning oxygen. Slapping on some drought-tolerant shallow plants, such as grasses, herbs, succulents, and mosses is no big deal, and won't require much maintenance. If you don't own your home and you're not going to be able to convince your landlord to convert the roof to a photosynthesizing wonderland, rooftop gardens have a similar effect and are easy to start and keep up. Rooftop gardens and mini-farms are rapidly gaining in popularity in big cities—some restaurants swear by their rooftop herbs and make cocktails from their little stashes of roof-grown flowers and fruits.

If you don't want your house to look like a total tree hugger's haven, you might consider a "cool" roof, which gets its name from the effect it has on the structure below—a cool roof is made of a reflective and relatively heat-resistant material, such as white vinyl, which can reflect more than 80 percent of solar rays. An existing roof can also be coated with a solar reflective, which is effective in reducing solar absorption. The Cool Roof Rating Council is a good place to start for more information: http://www.coolroofs.org.

Installing a water catchment system on the roof is also beneficial, and relatively simple. The water can be used straight for gardening or, for an extra investment, a filtration system can lead to showers from fresh rainwater.

In terms of smart building materials: Metal is great because it is completely recyclable; and multi-ply modified bitumen roofing systems are made of post-consumer crumb rubber from recycled tires.

I Need Some Facts to Bore My Friends With

Over 250 million tires are tossed into landfills every year in the United States, so a rubber roof is a no-brainer for the eco-conscious. Rubber is quickly becoming a commonly utilized material for building, since there's a serious surplus of used tires floating around, begging for a second or third life.

Out the Window: Windows and Weatherizing

Cut to the Chase, Hippie: What's the Least I Need to Know?

Drafty windows mean high energy bills: Up to 10 percent of your home's heat or air-conditioning can slip through the tiny cracks around a frame. Caulk and plug any leaks, and consider installing storm windows or covering panes with plastic sheeting in the winter.

Intriguing . . . I Can Handle a Little More

You could live in a cave, I guess, and not worry about windows. Or you could just reinforce them. Basically, you want to slap something on your existing windows to cut down on energy loss. See-through solar shades or reflective film will do the trick. Whatever you choose, just avoid PVC.

I Need Some Facts to Bore My Friends With

If you want to really get more serious, replace your windows with energy-efficient models. They aren't cheap, but some newer window designs can actually help heat your home in the winter. Even if it's 20 degrees outside, high-quality windows shouldn't feel cool to the touch, thanks to their U-factor, or the rate at which the glass conducts nonsolar heat. Double- or triple-paned windows should help to keep the panes at room temperature, but, like anything, they range in quality. The National Fenestration Rating Council (yes, they have organizations for *everything* these days) rates windows according to the efficiency of not just the glass and glazing but the frames and spacer materials.

Living Off the Grid

Cut to the Chase, Hippie: What's the Least I Need to Know?

It's not just for the Amish: Dropping off the grid is quickly becoming a workable alternative to dependency on government-provided resources—even in urban areas.

Intriguing . . . I Can Handle a Little More

Eschewing the grid means putting all your green elements together: You'll install solar or wind power and water catchment systems, and you'll produce so little waste there's no need for garbage service anymore. A family in California has gotten a lot of attention lately for being so dedicated to the zero-waste lifestyle that they accumulated only a one-liter glass jar of trash for the entirety of last year.

Off-grid living isn't just about going solar and letting rainfall or wells serve as your natural faucet. In the winter, fireplaces and wood-burning stoves replace heaters to keep the family warm, and come summer there's usually a small farm involved—planted on five acres or an urban rooftop if that's what you have to work with.

As for garbage removal, if you're like that family in California, who needs it? Off-grid living comes with its challenges, but over time you'll find that buying shampoo in disposable plastic bottles is overrated anyway. The zero-waste/off-grid combination means finding stores that sell amenities in bulk and bringing your own bags and containers to carry your purchases home. Off-grid living is certainly a simpler life, but it doesn't have to be depressing. I once saw a TV show where Ed Begley Jr. powered his toaster with an exercise bike—and that image could actually pull me out of a depression. If the thought of such a commitment is overwhelming, try baby steps: Convert to solar power, source at least part of your water from a catchment system, or rescue a goat.

I Need Some Facts to Bore My Friends With

It's not just private homes that are going rogue, but businesses and organizations, too: The 185,000-student Los Angeles City College is well on its way to becoming completely off the grid in a matter of years, perhaps with photovoltaic energy installed on a parking structure. It's *some* redemption for paving paradise.

who knew? "the garbage warrior"

Eco-architect Michael Reynolds, based in Taos, New Mexico, is dedicated to creating homes that are completely self-sufficient, and pretty breathtaking in their elegant design. Called "earthships," these structures are literally made from garbage—empty plastic bottles, beer cans, and tires; they utilize rain catchment and other devices to work with nature to provide an entirely green living environment. Finish that bottle of soda? Why throw it away when you can just wedge it into the wall?

While Reynolds' earthships first gained popularity in New Mexico, today they can be found all over the world, along with opportunities to join him and his team in their trash-using operation. There was recently a monthlong earthship and trash reclamation training in Haiti, and plans for the first urban earthships are under way in Philadelphia.

And by the way, what's a cooler pick-up line than saying you live in an earthship? Provided you just drop it in as a casual aside; you don't want to flaunt your earthship. I hate when people do that.

Warming Up and Cooling Off

Cut to the Chase, Hippie: What's the Least I Need to Know?

To save money on energy bills, you don't have to get rid of your furnace or boiler—just be sure your heating and cooling devices are running efficiently and don't go leaving them turned up or down when you don't need them. Or, you know, put a coat on.

Intriguing . . . I Can Handle a Little More

Since heating can account for about half of the average energy bill, it might be tempting to rush out and buy a newer model, but you can tame an old furnace with a few retrofitting tricks. Add a heat pump and seal your ducts to ensure that you're not losing heat before it even enters your home. If you're going to buy a new furnace, condensing furnaces and boilers are the most efficient: They capture heat from the exhaust and, instead of wastefully sending it up the chimney, convert it into steam and reroute the heat from the steam back into your toasty home.

If you live in a hot climate and need an air conditioner, make sure you replace it every eight years or so. A new system costs, but you'll ultimately save money in cooling bills. Also, changing air-conditioner filters regularly will increase performance.

I used to be such a cold junkie, I'd keep my thermostat at 64. My roommate at the time walked around bundled in winter clothes. I gradually nudged the thermostat up and realized that we acclimate to whatever temperature we're accustomed to. So nudge yourself closer to the temperature outside. You'll get used to it. Just ask my former roommate in the ski jacket.

I Need Some Facts to Bore My Friends With

The better insulated and the better sealed up your place is, the more efficient your heating and cooling is going to be. In Germany, where they like to brag about these things, there are some houses built so tight that a lightbulb can literally heat a room.

Humidifiers and Dehumidifiers: Boring but Necessary

Cut to the Chase, Hippie: What's the Least I Need to Know?

Let's talk humidifiers and dehumidifiers. I know, this has turned into quite the page-turner. But I'll try to make it interesting for you. Humans are delicate creatures: Super-dry climates bring chapped lips and cracked skin, but high humidity breeds mold and makes everyone's hair look a little too '80s. And not in a cool retro way.

Intriguing . . . I Can Handle a Little More

The recommended relative humidity for your home is between 30 and 50 percent. Humidity lower than 30 percent can dry out your respiratory system, leading to increased vulnerability to colds and chronic conditions like asthma. On the flip side, summertime's moist weather in many climates can lower your indoor air quality—and make you feel like you're swimming through soup.

When shopping for humidifiers and dehumidifiers, just like when shopping for other appliances, it's safe to fall back on the Energy Star label, slapped on the side of products that are more energy-efficient than most. This rating, combined with a good estimate regarding how much humidification or dehumidification you need, will save you about $20 a year in energy costs, or $230 over the course of a unit's lifetime.

I Need Some Facts to Bore My Friends With

Energy Star's website has a handy dehumidifier calculator to help you when it's time to figure out how much power your unit should have. And make sure it has a built-in humidistat, which allows you to choose your desired humidity level and programs the appliance to turn on and off automatically. Okay, so maybe there's no way to make humidifiers and dehumidifiers interesting. At least we're through it now.

Renovation: Things to Watch Out For Besides the Budget

Cut to the Chase, Hippie: What's the Least I Need to Know?

Keep it in the family, keep it local, and keep it looking new as long as possible. Rather than tossing out old lamps and importing new flooring from China, give away what you don't need and source new flooring from local or refurbished wood, or just refinish your floors using natural sealants and varnishes.

Intriguing . . . I Can Handle a Little More

Whether you're putting up a new wall or taking down old cabinets, choose low-emission, locally produced building materials. If it's time for new carpeting, don't sweat it: Choose carpets made from wool or recycled materials. And find an ecological company that doesn't treat carpets with toxic fire-retardants and other chemicals. The same goes for other replacements and improvements. Whether you want a new couch or a new television, hit local resale shops or online retailers like eBay before you buy a brand-new item. Likewise, think about recycling your old appliances, either by selling them or giving them away, so they don't rot in a landfill for the next thousand years.

I Need Some Facts to Bore My Friends With

I know the hell of renovation, and the last thing I want to do is add to your punch list, but as long as you're at the hardware store buying things anyway, you might as well look for "nontoxic" and "no VOC." While you're renovating, replace incandescent lightbulbs with compact fluorescent ones or LED lighting. You'll reduce your greenhouse emissions—and if everyone in this country switched five incandescent bulbs to compact fluorescents, the energy saved would be equivalent to taking 10 million cars off the roads.

Asbestos: Remedying a Bad Building Idea

Cut to the Chase, Hippie: What's the Least I Need to Know?

If you live in a house built before 1978, asbestos may lurk in your walls or under layers of paint. But before you knock down those walls, remember that asbestos is generally only toxic when it's disturbed, which is why home improvements and remodeling can be dangerous in older buildings.

Intriguing . . . I Can Handle a Little More

Before we knew its harms—or very much about it at all—asbestos became a leading ingredient in building materials because of its affordability and resistance to heat, fire, and electrical and chemical damage. A few years and many lung problems later, asbestos is a known carcinogen.

The most common type of asbestos is chrysotile, found in drywall, plaster, vinyl floor tiles, mill-board, caulk, paints, and other products. If you're considering an addition or a related improvement, have a professional get the job done. Abatement is affordable and investing in a professional is worth it. That way, you're not running the risk of exposing yourself to toxic chemicals and potentially cancer-causing particles.

On the plus side, a home inspector recently discovered asbestos near an air-conditioning unit in my childhood home. No big deal. I lived there only eighteen years.

I Need Some Facts to Bore My Friends With

There's no such thing as safe asbestos, but the EPA classifies a material as harmful if it contains more than 1 percent of asbestos. That said, many products can and do still legally contain trace amounts of it, which is like saying I just put a little arsenic in your drink. To figure out which products legally contain asbestos, refer to the EPA's clarification statement on their website: http://www.epa.gov/asbestos/pubs/asbbans2.pdf.

Solar Power: Because the Oil Companies Don't Own the Sun

Cut to the Chase, Hippie: What's the Least I Need to Know?

Coal, natural gas, and oil are responsible for powering most homes. None of these sources is renewable. If you can go just a little bit solar—a panel here, a strategically placed window there—you'll save yourself hundreds of dollars on energy bills and save us all from extra greenhouse gases.

Intriguing . . . I Can Handle a Little More

Active solar systems use solar panels and other technology to capture the sun's heat and convert it to electricity. Passive solar design, on the other hand, means tactical architecture—orienting a building to make use of the sun without installing mechanical devices or producing actual electricity. For example, a contractor might orient windows to maximize or minimize light and heat. Passive solar home design can also be more involved by incorporating a solarium into its architecture or adding overhangs and plants around a structure's perimeter to reduce the need for air-conditioning. Similarly, you can use building materials high in thermal mass when constructing rooms that will receive direct sunlight, and make the most of cross-ventilation by ensuring there are at least two openings (windows or doors to the outside) in each room.

Solar power can be applied to various appliances, too. Take the water heater: To go active, which is the most efficient move, be prepared to shell out some serious dollars. But a passive solar water heater uses convection (instead of pumps) to move water through your pipes. To go even greener, passive solar water heaters incorporate water catchment devices installed on the roof, which can then be distributed into pools, radiant floors, and other water needs.

I Need Some Facts to Bore My Friends With

Solar technology has been around since 1860, but wasn't seriously developed until the 1970s when oil prices began to rise and petroleum became less readily available. The White House itself once harvested the power of the sun. In 1979, the Carter administration installed thirty-two solar panels and used them to heat water. Seven years later, after gutting the research and development budgets for renewable energy, the Reagan administration quietly had the panels removed. Now they're housed in a science museum in China. Shhh . . . I won't tell if you won't.

who knew? frickin' fracking

Forcing fuel from the rocks of the earth puts the environment and our health in a tight squeeze. Fracking is the process of drilling deep into the earth to expel natural gas. For each "frack," up to 7 million gallons of water (along with sand and almost six hundred different types of chemicals) are blasted into a well, breaking the shale and creating cracks out of which natural gas flows. Most of this dirty water then mingles with the water table, contaminating our drinking water. Watchdogs are calling for stronger regulation of this industry, with zoning that avoids fragile ecological or residential areas.

Wind Power: Blow on This

Cut to the Chase, Hippie: What's the Least I Need to Know?

Wind power, solar power's less popular and more needy cousin, is easy to install and use to power your home—but only if you live on a few acres and don't mind clearing a space for the turbine.

Intriguing . . . I Can Handle a Little More

If you live in a part of the world with great open spaces, you can capture energy from the wind and convert it to electricity. Ideally, wind turbines should be at least 20 feet above any surrounding object and have at least a 250-foot-radius clearance around them. They're most commonly found in large families, called "wind farms," but you can install your own solo turbine if you've got the space.

Similar to solar energy, you'll have to part with a significant sum to buy and install a wind turbine, but after the initial investment, it's smooth sailing.

I Need Some Facts to Bore My Friends With

The technology behind wind power is pretty basic: The wind blows, two or three propeller-like blades turn around a rotor, the rotor is connected with a main shaft, and the main shaft spins a generator to power your television, microwave, and other power-hungry appliances. Like solar panels, wind-powered generators make DC electricity, which is then converted to AC with an inverter. If you connect your wind power to your area's electrical grid, you can net-meter your wind power; otherwise, you'll need batteries to store excess power for the less windy days. But if you generate extra wind power, some electric companies will buy the power from you, which makes this option all the more financially viable.

Check out the U.S. Department of Energy's Energy Efficiency and Renewable Energy website (http://www.eere.energy.gov/topics/wind.html) to find out whether your area can supply high-enough wind speeds to make the installation of a turbine worth your while. That website has a handy state-by-state guide that charts wind speeds across the country. Because, honestly, what would suck more than installing a wind turbine only to discover that there's not enough wind?

Lead and Other Reasons Not to Lick the Walls

Cut to the Chase, Hippie: What's the Least I Need to Know?

Asbestos isn't the only invisible carcinogen that could be skulking around in your walls. Lead-based paint often hangs out on window frames, walls, and the outer surfaces of homes built before 1978, when the federal government outlawed the stuff for good. Super-toxic lead pipes weren't banned until 1986.

Intriguing . . . I Can Handle a Little More

Lead is classified as a heavy metal, which means it doesn't break down, and just a few specks of lead dust can harm a child. Lead affects almost every part of the body, and potentially causes convulsions, comas, and, at high levels, even death. Lead in lower doses has been linked to problems within the brain, central nervous system, blood cells, and kidneys.

It most commonly leaches into the water system as water flows through lead pipes, or enters the airstream via decaying paint chips and deteriorating paint dust. This can lead to dangerous levels of it in both the water and air. It can happen naturally, or be exacerbated when lead-based paint is improperly removed by dry-scraping or sanding it.

If you live in an older home, you've got to do a little detective work. Check the outside of your house for flaking paint that might be lead based. Inside, look for dust that builds up around hinges and window frames or flakes that appear when two painted surfaces rub together, like when doors or drawers are opened and closed. If you see something suspicious, buy a home swab kit from your local hardware store and find out if you have a case of lead contamination. If you do have lead, consult a professional about your options for remediation.

When it comes to wallpaper adhesive, the easiest nontoxic option is to make your own. Combine one cup of flour and three teaspoons of alum in a double boiler, add enough water to make it the consistency of heavy cream, heat it up, stirring constantly until it thickens to a more gravy-like texture. As it cools, stir in ten drops of clove oil. Apply it with a glue brush. You can keep this stuff in a glass jar with a screw top in the refrigerator for up to a couple of weeks.

I Need Some Facts to Bore My Friends With

Van Gogh's insanity, some historians say, could have been due to lead poisoning—all that lead paint he inhaled might have contributed to, say, the brilliant decision to cut off his own ear.

I'm Donald Trump

If you suspect there's lead in your walls or pipes, hire experts and let them deal with the dirty work. The EPA's "Renovation, Repair and Painting Rule" requires that renovation firms that carry out repairs that disturb lead-based paint and pipes be certified by the EPA, which means they're reliable and knowledgeable in lead-safe practices. Check out the National Lead Hotline for a list of contacts in your area: http://epa.gov/lead/pubs/nlic.htm.

Okay, I've Got My Own Place, but I've Also Got Credit Card Debt

Go slowly: It's okay to gradually upgrade your home by replacing doors, window frames, paneling, or drywall over the course of several years, as you can afford to do it.

I'm Sleeping on My Friend's Couch and Eating Ramen Noodles

Lead poisoning, like so many health problems, is more common when we're living in poverty. An estimated 85 percent of lead-poisoned children are eligible for Medicaid, so use any health-care resources you can and get yourself and your kids checked out.

Jump In: Nontoxic Pools

Cut to the Chase, Hippie: What's the Least I Need to Know?

If the itchy eyes and green-tinged hair that are side effects of chlorine aren't enough to turn you off, the toxic chemical has been linked to asthma in children and other diseases. With a little foresight, the pool can be a place of both refreshing *and* nontoxic relaxation.

Intriguing . . . I Can Handle a Little More

Pools need to be cleaned and disinfected to be safe and appealing environments for swimming, and alternatives to chlorine-based solutions abound. The methods vary, including saltwater pools, cycled with sodium chloride, which contain low levels of chlorine due to the outgassing of the salt but are much gentler and safer than chlorine-cleaned pools; ionization pools, which use copper and/or silver ions and very little or no chlorine or salt; oxidation pools, which employ ultraviolet light or electricity via an electrical generator; sonic cleaning, which sends ultrasonic waves through the water to take out algae and other contaminants on a cellular level; oxygenation, which conditions and restructures water through the use of a special water filter and hydrogen peroxide; and ecosystems, which are full of plants and have a breathable bottom, mimicking the natural cleaning systems of lakes or ponds.

Chlorine unsurprisingly has negative effects on the earth as well. It's bad for precious organisms living in water and soil, and has been implicated in the depletion of the ozone layer, global warming, and acid rain.

I Need Some Facts to Bore My Friends With

According to the Environmental Protection Agency, exposure to high levels of chlorine can be hard on the respiratory system in addition to irritating the skin and eyes. Repeat exposure to chlorine seems to be harmful to the immune system, the blood, and the heart as well. While the average swimmer won't be dealing with these types of extremely high levels of chlorine, it's best to skip it altogether if possible. It's not just about health or the environment, either: other methods, such as oxygenation, are much cheaper than hiring pool technicians or buying expensive, generic pool-cleaning chemicals. Healthier bodies and thicker wallets make the avoidance of chlorine a no-brainer.

Clearing the Deck

Cut to the Chase, Hippie: What's the Least I Need to Know?

By this point, you've gotten the idea: Pretty much everything on the wide green earth can be gentler to you and the environment with a little brain work. This, of course, applies to decks and terraces just as much as it does any other building material. So choose your wood mindfully and your sealant carefully.

Intriguing . . . I Can Handle a Little More

Similar to play equipment, decks must be made with sustainable wood, such as cedar, Forest Stewardship Council–approved, and finished with water-based or other natural sealant to be truly environmentally savvy. It's also important that the manufacturer uses minimal fossil fuels to cut and treat the wood. Locally sourced wood is great, and reclaimed wood is even better. Some timber outlets pay their sources fairly and make sure that another tree or two is planted for every one cut down; do a little research and stick to those outfits.

Every other year, you'll have to reseal your deck to prevent warping and wear. Alkaline copper quat (ACQ) is a popular choice of preservative; while some research has been done and the chemical components of ACQ were not associated with adverse effects in humans, it's still a chemical and thus might not be the absolute safest option. Alternately, Earthpaint makes a nontoxic "Rainforest Sealer." You can also just use paint. It's an alternative to sealant, and there are nontoxic paint companies that offer beautiful colors made from clay, milk, and other all-natural ingredients.

I Need Some Facts to Bore My Friends With

The deck itself is a great start, but don't forget the planet when furnishing your patio. The safest route is to be sure everything is Forest Stewardship Council–okayed or made from recycled plastic. Sustainable deck chairs and other furniture is often made of durable eucalyptus, a great choice. Recycled plastic pieces are usually cheaper than wooden options, but double-check that these don't contain any PVC plastic, which can be harmful.

clean home

I ask people why they have deer heads on their walls. They always say because it's such a beautiful animal . . . I think my mother is attractive, but I have photographs of her.

—Ellen DeGeneres, talk show host and vegan

Change Begins at Home

Changes that you or I can make at home often come down to one simple choice that will benefit us and our kids for years. For example, if you just switch your paint to the zero-VOC kind, you can affect the quality of the air that your family breathes for a long time. The same goes for cleaners and furniture. This section is filled with little changes that might feel like a pain at first—even to me sometimes—but when you make these shifts, they quickly become habits.

If we could see the toxins in everything—if the carpet glowed neon yellow or the countertop had "cancer" written all over it—we'd make cleaner choices. As you read this section, start picturing bright warnings on your questionable interiors. That's what I do. Cheerful, isn't it?

Buying Furniture: Why Your Couch Is the Enemy

Cut to the Chase, Hippie: What's the Least I Need to Know?

One day some genius in California decided that couches should be able to withstand a flame for up to twelve seconds without catching fire. Unfortunately this idea had a nationwide influence, and it's quite likely that every day you are breathing in chemicals designed to keep your couch from bursting into flames. I know, you may be thinking, *But what if that helps to keep my house from burning down?* Well your couch will still ignite, just on a brief delay, and it will let off toxic gasses as it goes up in smoke.

Intriguing . . . I Can Handle a Little More

Based on a study by UC Berkeley and Duke University, whether or not you live in California, your living room couch is probably off-gassing all kinds of toxic stuff. Studies (including the one mentioned above) have found that while your couch may be stain- and fire-resistant, the substances reducing its flammability and stainability can lead to severe health problems—anything from cancer to lowered IQ to decreased fertility. Although the commonly used PBDEs (which may be as dangerous as DDT) were banned in 2005, the tenacious toxin still remains in our systems for more than twelve years after exposure, and worse, PBDEs can be found in human breast milk, newborns, and even fetuses. And a lot of you still have that ten-year-old couch. But even if you bought your couch after 2005, all that really changed was manufacturers replaced one fire-preventing chemical with another, most likely chlorinated Tris or Firemaster 550. Chlorinated Tris is known to mutate the DNA of people exposed to it. That doesn't sound like a big deal to me. (Read sarcasm here.)

As for Firemaster 550, in spite of its cool action-hero name it is shown to be an endocrine disruptor in animals and who-knows-what in humans. But I'm sure it's fine. (Again, read sarcasm here.) Google "green or eco furniture" and find a company in your area that can make you a sofa from organic cotton and wool and have them hold the chemical cocktail while they're at it. More and more companies are popping up that can do so on a budget.

Now while you're busy plotting your couch's exit from your house, don't forget about your cabinets, dressers, bed frames, closet doors, etc. Unfortunately, most wooden furniture is not actually wood at all anymore. To lower costs, manufacturers use wood veneer over a particleboard core. So even if you're pretty sure the face of your furniture is solid wood, look carefully at drawer bottoms and backings, where manufacturers are even more likely to sneak in the cheap stuff. Particleboard is almost always loaded with formaldehyde, a known carcinogen with no safe level of exposure. In fact, in spite of it also being linked to childhood asthma and allergies, it may be found lurking in your changing table or crib. Look for furniture made with real, sustainably harvested wood, sealed with a wax- or water- rather than oil-based finish, whenever you can.

I Need Some Facts to Bore My Friends With

Remember the FEMA controversy from 2008 in the Gulf Coast? If not, I'll remind you. It was discovered that the victims of the 2005 hurricanes Katrina and Rita had been housed in trailers that averaged a level of formaldehyde as much as five times higher than what is found in most new homes. Residents complained of bloody noses, difficulty breathing, and coughing, among other symptoms. In September of 2012, a federal judge approved a $42.6 million dollar class-action settlement against the makers and installers of the trailers in favor of affected residents.

Sleep on It: Bedding

Cut to the Chase, Hippie: What's the Least I Need to Know?

We spend about a third of our lives in bed, so if you can green only one thing at a time, start with where you and your family members are sleeping. Traditional bedding and mattresses use cotton that is treated with all sorts of cancer-causing chemicals and toxic fire retardants. Sounds like a nightmare to me.

Intriguing . . . I Can Handle a Little More

Bedding made of organic, nontoxic materials and dyed with plant- or clay-based dyes are not only safer but often softer and cozier, thanks to their eyes-wide-open manufacturers. Comforters and pillows stuffed with down and feathers are best in terms of low-carbon footprints compared to other fillers, but they're not cruelty-free, since they're sourced from ducks and geese that kind of wanted to keep their feathers to stay warm themselves. Choose from the million other all-natural fillers and fabrics, such as millet, buckwheat, hemp, or even green tea extract. Sounds cozy, doesn't it?

Once you've tackled bedding, try buying a green mattress. There are mattresses spun from organic cotton and wool, whose steel springs have no toxic chemicals painted on; foam-based mattresses made from rubber tree–based latex; and coconut husk–fiber mattresses. Additives range from aloe vera to castor oil.

I Need Some Facts to Bore My Friends With

The natural-bedding universe includes so many types of fabric and fillers that it can get a little overwhelming. For example, you can purchase a pillow filled with fibers made from the kapok seed, a naturally hypoallergenic little powerhouse that provides an alternative to down for those who are allergic or super-vegan.

Can't I Just Wash the Toxins Off with Soap or Sanitizer?

Cut to the Chase, Hippie: What's the Least I Need to Know?

I know you like your hand sanitizer and it makes you feel clean, but it can cause more trouble than it prevents. All soap will get your hands and body clean, but some leave behind more chemicals than they wash off.

Intriguing . . . I Can Handle a Little More

In 2002, scientists in Germany performed a study that concluded hand sanitizers can be harmful to children. In 2008, a similar study was done in the United States, and it echoed the conclusions of the European study done six years before. Sure, these antibacterial substances kill germs, but they don't discriminate, so friendly bacteria are offed, too. What's more, most of these sanitizers are heavily infused with alcohol, which can produce biofilms and lead to the production of pathogens, such as staph. Biofilms are nearly impermeable layers that can form in deep layers of skin, which can then cultivate strong strains of bacteria that even antibiotics can't get through. Enough to make me want to not wash my hands.

Many shampoos, body washes, and other soapy body products also include alcohol. The best place to shop for soap is the health food store. You'll find natural soaps infused with essential oils such as rose or geranium. They smell great and clean as well—if not better—than the stuff you can buy at big box stores.

I Need Some Facts to Bore My Friends With

For an all-natural hand sanitizer, mix essential oils of lavender, peppermint, eucalyptus, clove bud, and/or thyme with olive oil and rub it on your hands. Or just wash with plain old plant-based soap and water. Not only will your hands feel and smell better than they would have after the use of store-bought sanitizer, your health won't suffer: One study found that people who use alcohol-based sanitizers contract colds and flus at the same rate as people who wash the old-fashioned way. Some illnesses, like the norovirus, a common stomach flu, is resistant to most hand sanitizers anyway.

Paint It Black

Cut to the Chase, Hippie: What's the Least I Need to Know?

Because of the paint we use, the EPA ranks indoor air quality among the top five health risks. Scan your paint label for "Low-" or better yet, "Zero-VOC"—that's volatile organic compound to you . . . but it's not the good kind of organic. VOCs include cancer-causing toxic contaminants.

Intriguing . . . I Can Handle a Little More

Unless you're mashing berries and red rocks, there's almost always going to be some uninvited ingredient in your house paint. But long-term exposure to VOCs, whose gasses can be emitted in either liquid or solid forms, can instigate asthma attacks, nausea, and dizziness, and cause long-term kidney damage.

For the chemical-conscious home, there's a sizable list of natural and even vegan paint manufacturers who draw ingredients strictly from food or mineral-based sources: Devine Color Paint, Ecos Paints, and Earthborn Paints. These paint peddlers look to resins, plant oils, chalk, talcum, beeswax, and other recognizable ingredients to create rich colors for human homes. Look for the term "Fully Disclosed Ingredients" because, believe it or not, the companies that make that distinction are the only ones legally bound to tell you what's really in that can.

I Need Some Facts to Bore My Friends With

Even a zero-VOC formula could still contain up to 5 percent formaldehyde or any other variety of naturally occurring but noxious additives. While low-VOC formulas can still contain fungicides, biocides, or abrasive colorants at reduced levels, these items are not yet fiercely evaluated by purveyors of air quality.

Then again, I can just send my kids over with a bucket of pomegranates and your walls will be painted nontoxic style in no time.

---

who knew? greenwashing

Dubbing a product "sustainable" is hot. But just because a product has a picture of a tree on its label doesn't mean it's green. There are two types of "greenwashing": Companies might try to sell something by claiming it's environmentally sound, or they might want to cover up for something they did that's bad for the planet by using earth-friendly distraction tactics. One such smokescreen: a company that claims it's green because it offers paperless billing while producing tons of wasteful bill-stuffers, newspaper inserts, and billboard ads. To find the truth about a company or product, visit http://www.greenwashingindex.com.

Pest Control: Because Sometimes Asking Mice to Leave Nicely Doesn't Work

Cut to the Chase, Hippie: What's the Least I Need to Know?

If you've got a bug outbreak, but no time to concoct a potion, you can find highly effective eco-friendly aerosols at your local health-food store. All the ingredients should be recognizable by their common names, like cinnamon and rosemary oil—so natural you could practically eat it. But don't. Unless you have tapeworm. Okay, still don't eat it.

Intriguing . . . I Can Handle a Little More

Mice, cockroaches, house centipedes, sugar ants, overgorged flies infest mansions and basement apartments alike. But petrochemical pesticides designed to kill pests aren't good for people either. Ninety percent of American households still use some form of caustic bug-killer, which put its inhabitants—especially kids and pregnant women—at risk for side effects, including long-term damage to the liver and the central nervous system and increased risk of cancer.

Spiders, in fact, are far better friends than pesticides—they eat our fruit flies without asking for anything more than a shady corner to call their own. But if you don't want to share your pad with Charlotte, essential oils work as well as poisons. Most pests can't stand the likes of the mint family and just take off. Keep a spray bottle of peppermint, rosemary, tea tree, and a touch of mineral oil. It can knock out a roach, mouse, or palmetto bug problem in a flash. Flies will steer clear of sachets of crushed mint leaves. On the same note, after completing a routine housecleaning, douse a cloth in any one of those minty oils and wipe around the edges of the room, the counters, the base of refrigerators, and other nooks where critters like to hang out.

I Need Some Facts to Bore My Friends With

Now, I'll trap a spider and carry it outside, but most won't. If you're less green—like the people in my life who want their bugs dead as doornails—try diatomaceous earth. It's a common eco-friendly insecticide—a naturally occurring sedimentary rock that is crushed into a fine, white powder and sprinkled in bug zones. Insects lap it off their skin and eventually dehydrate to the point of losing their exoskeleton. Okay, maybe that counts as animal cruelty. But it's been used for hundreds of years in everything from killing parasites in zoo animals to filtering public swimming pools to stabilizing dynamite. So check your internal moral compass and decide for yourself how you feel about ending a fly's life.

Bathrooms: Money Like Water

Cut to the Chase, Hippie: What's the Least I Need to Know?

Twenty-six percent of the water used in the average home is just flushed down the toilet. At a minimum, don't use your toilet as an ashtray or wastebasket—every time you flush a butt or a tissue, that's seven gallons of water down the pipes.

Intriguing . . . I Can Handle a Little More

You can start saving water—and money—in the bathroom with very little investment. To reduce water wasted down the toilet, get two quart-size plastic bottles with screw-on caps and fill each with about an inch or two of sand or pebbles to weigh them down. Fill the bottles the rest of the way up with water, then screw the caps on and put them into your toilet tank, safely away from the operating mechanisms. If you've got kids, I guarantee they'll want to help out with this because it will give them an excuse to fill your toilet with sand and pebbles. You could also buy a "tank bank" for a couple of dollars online instead. Just be sure that at least three gallons of water remain in the tank so it will flush properly. This simple strategy serves to reduce the amount of water in your tank by about .8 gallons, saving that amount of water with each flush.

Next, pick up a water-saving low-flow shower-head or restrictor. Even I can install one of these things.

Save additional water by limiting your showers to the time it takes to soap up, wash down, and rinse off.

When you've got a little more time and money, go ahead and replace your old toilet with an eco-friendly one. Frankly, *any* new toilet will use less water than an older model.

I Need Some Facts to Bore My Friends With

If you're among the one-quarter to one-third of the U.S. population relying on a septic tank, prepare to be grossed out. Flushed household chemicals and sewage are poisoning the groundwater. An EPA study of chemicals in septic tanks found toluene, methylene chloride, benzene, chloroform, and other volatile synthetic organic compounds related to home chemical use, many of them cancer causing.

Between 820 and 1,460 *billion* gallons of this contaminated water are discharged to our shallowest aquifers each year. In fact, septic tanks are reported as a source of groundwater contamination more than any other source. The word "septic" comes from the Greek *septikos* which means "to make putrid." Sexy, isn't it?

And the Hippie Nerd of the Century Award Goes to . . .

For a few hundred to a few thousand dollars, you can invest in a "dry toilet" and central waterless composting system. Yep, no water—and you use your poop to fertilize your garden. For the record, I'm not doing this. Yet.

Don't Flush That Pill: Pharmaceuticals in Wastewater

Cut to the Chase, Hippie: What's the Least I Need to Know?

The FDA actually advises consumers to flush hazardous and excess pharmaceuticals, but by doing so we've helped poison our septic systems and groundwater supply. The next time you have a bunch of pills to get rid of, look for state-approved community drug take-back programs in your area. Your local trash and recycling services will have access to schedules—and you can tape them on your medicine cabinet if you're too stoned to remember.

Intriguing . . . I Can Handle a Little More

Since the early 1990s, traces of over a hundred pharmaceuticals have been showing up in American drinking water. While the usual suspects are to blame (pharmaceutical companies and hospitals, in this case), household flushers are high contributors. The resulting groundwater toxicity is especially hard on children. Studies show that American kids now regularly display small amounts of unprescribed antibiotics and lithium in their blood. Fish and strains of oxygen-giving plankton fall prey to this waste, too, and it damages regenerative and reproductive systems.

I Need Some Facts to Bore My Friends With

In 2008, the DEA started sponsoring annual National Prescription Drug Take-Back Days in all fifty states, plus American territories. The 2011 collection yielded 377,086 pounds of unwanted or expired medications that were handed over for safe and proper disposal. I've had some friends who would "take back" your drugs for you, too. But that's another book.

Hippie Household Cleansers

Cut to the Chase, Hippie: What's the Least I Need to Know?

You can clean virtually anything in your home with an easy combination of equal parts white vinegar and warm water in a spray bottle.

Intriguing . . . I Can Handle a Little More

For a streak-free glass cleaner, use a two-bottle attack plan. In your first bottle, mix two tablespoons of rubbing alcohol with a quart of water. Your second spray bottle is just your all-purpose white vinegar and warm water solution. Using a lint-free cloth or newspaper, coat the window with the rubbing alcohol mixture. Now spray it with the vinegar solution and use your cloth or newspaper to clean the glass just like you would with your blue toxic window cleaner.

For dirty wood floors that need more than sweeping, damp-mop unvarnished floors with liquid Castile soap diluted in two gallons of warm water. For varnished or no-wax wood floors, damp-mop with a solution of one part white vinegar to ten parts warm water.

For ultra-greasy windows your kids have smashed their fried tempeh into, add a half tea-spoon of clear liquid soap to the vinegar solution. Tired from all this cleaning? Yeah, me too. And I'm just writing about it.

I Need Some Facts to Bore My Friends With

There are so many potential hazards in store-bought cleaners that if you're not going to make your own, at least head to the more green markets and read the labels before you buy them. The range of chemicals in standard cleaners have been linked to serious injuries and illnesses, including blood disorders, liver and kidney damage, reproductive and nervous system harm, depression, and cancer. There are tons of toxic chemicals to avoid but a few to look out for are ethylene-based glycol ethers, which are considered hazardous air pollutants under the U.S. Environmental Protection Agency's 1990 Clean Air Act Amendments; phenols; and butylcellosolve, which is commonly found in all-purpose and window cleaners (and should win a prize for the chemical with the most far-reaching damage, as it harms the bone marrow, kidneys, liver, and nervous system).

And by the way, like that pine or lemony smell in many products? That's thanks to terpenes, which react with ozone to produce toxic compounds. Ummm . . . call me crazy, but I'll stick with lemon.

Dishing It Out: Your Dishwasher Just Killed That Fish

Cut to the Chase, Hippie: What's the Least I Need to Know?

Unless you want to eat with your hands like the ancient Romans—or my kids for that matter—you're going to have to wash some stuff. Running a dishwasher is greener and more cost-effective than doing the dishes by hand—but only if you're running full loads. Choose dishwashing soap that's phosphate free. While phosphates are not directly harmful to humans, they are harmful to the environment.

Intriguing . . . I Can Handle a Little More

Phosphates are inorganic additives that deter the unsightly effects of hard water minerals famous for leaving blemishes or browning on dishes and clothing. And while phosphates have been banned from laundry detergents in the United States since the mid-1990s, dishwashing soaps (for dishwashers) have managed to escape scrutiny. Phosphates are cut from the same kingdom of chemicals used in commercial agriculture fertilizers—not stuff you want to cuddle up with or eat off of. The phos-

phate runoff from our laundry-obsessed culture has caused all kinds of algae to overrun aquatic ecosystems from coast to coast. Basically, the phosphates from our detergents get into our lakes, and algae grows on phosphates. I've got nothing against a little algae, but excessive algae growth causes a reduction of light and oxygen, ultimately making the water more acidic. When aquatic plants do not get the light they need in order to complete photosynthesis, they die. Fish that feed on those plants—and on the oxygen in the water—die, too. If you are heartless when it comes to the well-being of fish, at least avoid soaps with petroleum, because in large amounts it is actually toxic to us.

I Need Some Facts to Bore My Friends With

You can make your own powdered dishwashing detergent using ingredients you already know the names of. Mix a cup of Borax and a cup of baking soda into a 16-ounce container, and shake it with the top on. That's it. You'll need only two tablespoons per load, and a finger of white vinegar in the rinse module should wipe out any residue. If you're attached to the olfactory experience of cleanliness, add a drop or two of lemon, eucalyptus, or tea tree oil.

Cleaning the Furniture: Sit on This

Cut to the Chase, Hippie: What's the Least I Need to Know?

If you're like me and you spill beet juice or pomegranate pulp on something expensive or antique every week, remember there's only one cleaner that costs less than a dollar a gallon and won't kill the fish: white vinegar.

Intriguing . . . I Can Handle a Little More

Mixed with a little bit of bottom-shelf olive oil for leather or plain old water for plant fibers, vinegar can wipe out just about any stain. Put the kibosh on harsh chemicals, which can seep into your skin and clothes, in a setting where you should just be able to relax.

Add a tablespoon of vegetable oil to a jar of clear vinegar, give it a shake, and run a soaked rag over the surface of your couch or chair to restore leather's righteous sheen. The same solution goes for any fine wood pieces. Three parts vinegar, one part water will lift even the deepest dirt streaks from cottons, linens, and even synthetic blends. For really serious stainage, dab a bit of 3-percent hydrogen peroxide on the site.

I Need Some Facts to Bore My Friends With

Most commercial brands of upholstery and carpet cleaners contain perchlorethylene, a carcinogen, which, along with many other side effects, can cause respiratory upsets even for those in the best of health.

Books, Newspapers, and Magazines: Worth the Dead Trees They're Printed On?

Cut to the Chase, Hippie: What's the Least I Need to Know?

If you read fewer than twenty-three books a year, owning a Kindle or e-reader won't actually offset your carbon footprint. Go to the library or share your books with friends.

Intriguing . . . I Can Handle a Little More

If you read more than a couple of books a month, an e-reader is one solution to the paper industry's unstoppable waste. Between periodicals, newsletters, Harlequin romances, textbooks, Pulitzer Prize–winning novels, and every rag, well, some 125 million trees are being harvested each year for production. To make matters worse, many publishing houses still overprint—if a book doesn't end up selling well within the first few weeks, thousands of copies are pulped, sending more fossil fuel into the atmosphere.

I Need Some Facts to Bore My Friends With

Even if you read a lot, e-readers aren't a magic pill. Where trees are the vital resource used in paper production, a huge amount of water goes into electronics manufacturing—that's in addition to the fuel, fumes, and energy that follow these objects to the end of an often ineffective recycling road. Plastics and corrosive chemicals and the energy spent to keep your reader charged are often nonrenewable, and ultimately harder on our pocketbooks and the earth than old-fashioned paper.

The greenest option? Yep, go ahead and pay off those library fines and get back in touch with the Dewey Decimal System.

Dog and Cat Care: Because Fido and Mittens Want to Be Green, Too

Cut to the Chase, Hippie: What's the Least I Need to Know?

The beginning of becoming a green, clean pet owner is cheaper than going to the pet store and buying that mistreated little purebred: Get a recycled animal from your local shelter.

Intriguing . . . I Can Handle a Little More

Once you've rescued your new pet, you'll have to feed it. Swap out that big name commercial pet food—which is basically reconstituted slaughterhouse by-product, corn, and gluten—for a USDA-certified organic brand of kibble or kitty chow.

Or forget the fancy stuff and make your own. A kitchen concoction of beans, greens, eggs, and simple starches like quinoa, rice, or oatmeal provides pets with essential nutrients. Consult an animal cookbook or trusted online recipe resource like No Cans: http://www.nocans.com.

Dogs and cats can all be vegetarian or vegan. Consider Bramble, a twenty-seven-year-old border collie whose vegan diet of rice, lentils, and organic vegetables earned her consideration by the *Guinness Book of World Records* in 2002 as the world's oldest living dog. Even if you're not trying to set any longevity records, the nutritional needs of dogs and cats are easily met with a balanced vegan diet. Check out *Vegetarian Cats & Dogs* by James Peden.

Most holistic pet stores can also help you start your feline or canine on a raw meat diet if you prefer, with cuts from your local farmer's market, and help your pets to eat in a way they might in the wild.

When it comes to flea control, try lavender instead of toxic, store-bought remedies. ? fifteen drops of lavender oil with water ... spray bottle, then spray your pets with it. The lavender will act as a repellent and help heal existing bites.

I Need Some Facts to Bore My Friends With

American dogs and cats produce 10 million tons of waste each year. When you throw that poop in a plastic bag and into the trash, it becomes a nearly permanent part of the landfill landscape. I know you've been dying for one of these . . . go ahead and invest in a dog-doo/cat-poo composter instead. Or move to San Francisco—it's become the first city in the United States to consider converting pet feces into methane that can be used for fuel.

I'm Donald Trump

Source your toys from designer paw shops, where you'll find all manner of hemp, organic cotton, and recycled toys and beds. Make appointments with a naturopathic veterinarian, and invest in annual animal reiki.

Okay, I've Got My Own Place, but I've Also Got Credit Card Debt

If you're living in an apartment, plant an indoor lawn next to your sunniest window. Dogs and cats love to munch on the green stuff for digestive upsets and general gastronomical maintenance. If you don't have a common green, growing your pets their own private patch of wheatgrass is a minor expense, and keeps them happy, healthy, and pesticide free.

I'm Sleeping on My Friend's Couch and Eating Ramen Noodles

Become a dog walker and make cheap green pet food for your clients. By the time you're ready to be a pet owner, you'll be a pro.

Dirty Laundry: The One-Page Version

Cut to the Chase, Hippie: What's the Least I Need to Know?

Most of the energy spent doing a load of laundry goes into heating up the water. Save money and reduce carbon emissions by washing with cold water when you can.

Be choosy when it comes to detergent, too. Conventional, chemical-filled detergent affects our health—not to mention the health of plants and aquatic life in the waterways downstream. Look for detergents that spell out what's *not* in them. The longer the list, the better.

Intriguing . . . I Can Handle a Little More

When you're ready for a new washing machine, choose an efficient Energy Star unit even if it costs a little more up front. As for what you are washing with, conventional detergents often contain toxic and potentially cancer-causing chemicals. Some to look out for include sodium lauryl sulfate, sodium laureth sulfate, 1, 4-dioxane, and nonylphenol ethoxylate, which breaks down into more toxic compounds over time. They were smart to ban that last one in Europe. One more reason to move to Rome.

I Need Some Facts to Bore My Friends With

The U.S. Department of Energy estimates that 12 percent of the average household's utility bill—and carbon footprint—comes from heating water. I'm not ready to go back to a boiling pot in the fireplace, but I wash my laundry in cold water. Washing even half of our laundry in cold water cuts some 140 pounds of CO_2 emissions each year—not to mention cutting my utility bill by about 5 percent.

Washing machines are the second-biggest water hogs in the home—just after toilets. Investing in a high-efficiency machine means using 50 percent less water and up to 60 percent less electricity than an old-school top-loader. That saving can add up fast. If even 1 percent of U.S. households switched to water-efficient appliances, we'd reduce greenhouse gas emissions by 75,000 tons.

There you have it. Now, if you still feel like reading a whole book on green laundry, I'll send you my copy.

Millionaire-in-the-Making Bonus

The average American household can save an additional $100 a year on their electricity bill by being a little bit old-fashioned. Just abandon your dryer and hang clothes up on the line like Great-grandma used to do.

Clean Linens: The Cheat Sheet

Cut to the Chase, Hippie: What's the Least I Need to Know?

Before you leap for the bleach bucket, try a search-and-destroy method on your white linens by running a bit of cold water over the trouble areas (stains from blood, sweat, or dirt, for example), then dab on some baking soda and castile soap. Rub the stained parts of fabric together, creating friction that will loosen the gunk, then toss it in the machine.

Intriguing . . . I Can Handle a Little More

Even a green household likes to keep its whites from looking dingy, so keep the water cold and boost your repertoire of natural whiteners.

If the castile soap doesn't work, use strong lemon juice and a rubbing rag on stains and spread the laundry out under the sun (just like the old hair highlights trick—except it won't turn your sheets a weird bronzy orange). If you don't have the time to address sheet by sheet, throw a quarter cup of lemon juice into the load with half the usual amount of your trusted biodegradable detergent. For a softer finish, add a cup of vinegar.

I Need Some Facts to Bore My Friends With

When used correctly—and kept out of the reach of children and adults acting like children—chlorine bleach isn't super-dangerous: Bleach biodegrades almost completely into oxygen, salt, and water in the environment. But when mixed with other compounds, it can get dangerous. Never mix bleach with ammonia or acids—including vinegar. Unless toxic gas is your thing. But I'm guessing it's not if you're reading about cleaning linens in a green book.

Televisions: Yes, This Hippie Has One

Cut to the Chase, Hippie: What's the Least I Need to Know?

Keep your TV, and all boob-tube accessories (DVD player, gaming devices etc.), plugged into a single power strip so that everything can be turned off at once when not in use.

Intriguing . . . I Can Handle a Little More

Most TVs can rival even your fridge in an energy-drinking contest, especially the old tube models, which still lurk around a few old-school American living rooms. Still, unless your old tube is on the fritz, opt for some simple changes, like limiting the weekly usage, turning down the contrast, and unplugging the TV at night.

If you're in the market for a new television, keep it small and stay away from plasma. The larger the screen, the more environmentally costly it is to run. Keep your eyes peeled for ones that are mercury- and lead-free and don't use noxious flame retardants that endlessly release ether into your TV room. A liquid crystal display (LCD) screen is almost always the least detrimental in terms of energy costs, but also look for machines with "power-saving" modes, back-lighting, and re-cycled materials. Phillips, Samsung, and Sharp all have the highest-rated models for these needs, and often ship in postconsumer packaging, to boot. Another alternative is a rear projection TV set. It requires that you have a blank wall space available to be the screen, but it's got the lowest wattage of all.

I Need Some Facts to Bore My Friends With

Most electronics stores catch consumers by dis-playing their televisions on high-contrast settings, but what most people don't know is that you can save up to 30 percent of your TV energy use by turning down the brights. Most sets come with "showroom" and "home" modes that you can change with the click of a button, but if yours isn't that sophisticated, just lower the contrast and keep the lights off. Not only is watching TV in the dark more romantic, so is that lower energy bill.

who knew? geothermal and hydroelectricity

Geothermal energy has been around since Pa-leolithic times. It has enormous potential to gen-erate electricity, but still requires more research and efficiency. Just three miles beneath our feet there are hot spots filled with steam that can be directed into turbines to generate hydroelectric-ity. Then again, drilling for geothermal is an in-volved and expensive process. Some countries have figured it out, though: While geothermal energy accounts for only 1 percent of the world's electricity supply, in Iceland almost 85 percent of homes are heated geothermally, making Iceland the real green-land.

Computer Geek

Cut to the Chase, Hippie: What's the Least I Need to Know?

Next time you're on the prowl for a new computer, opt for a slim notebook or laptop with dual-core processing, a whole lot of memory, and extensive battery life. The less you're having to charge the thing, the better.

Intriguing . . . I Can Handle a Little More

Desktops typically consume five times as much energy as laptops, so there's no need to cling to the bygone era of the common living room computer. Unless you're a designer or a filmmaker, recycling your household desktop and replacing it with a laptop is the first step toward a greener cyber life. If you need a new laptop but are on a small budget, look into high-resolution netbooks, which tend to cost half as much as regular laptops, have extremely durable batteries, and generally the only sacrifice is no CD drive and less memory. No matter what laptop you choose, if you do more than a little keyboarding, remember to set up an ergonomical workstation. Hand cramps and repetitive stress injuries are more of a concern on compact keyboards. And, "I got this repetitive stress injury on my compact keyboard" is not a great conversation starter.

I Need Some Facts to Bore My Friends With

The EPA has developed the Electronic Product Environmental Assessment Tool (http://www.epeat.net) to help determine which computer models are produced without the use of mercury, hexavalent chromium, and noxious flame retardents you don't want to be breathing anywhere near. Look for the Energy Star to get started, which, among other things, promises a low-powered sleep mode.

If all personal computers in the United States followed Energy Star regulations, the reduction in greenhouse gas emissions would be equivalent to getting 2 million cars off the road.

Call Me: Cell Phones vs. Landlines

Cut to the Chase, Hippie: What's the Least I Need to Know?

While both types of telephones consume their fair share of energy, you can always choose to consolidate—ditch one of your lines.

Intriguing . . . I Can Handle a Little More

Landlines are inexpensive to run, but they require constant electricity. Cell phones can stay charged with only an hour or two of plugged-in time every couple of days—*if* we bother to unplug our chargers in the interim. Landlines are about as harmless as transistor radios, but as hard to recycle as computers. Cell phones use nasty, corrosive batteries, prompting recycling initiatives to be instituted nationally. Some sources recommend landlines for the eco-conscious caller, if only to conserve energy you'd otherwise spend worrying about brain cancer. But I'm so addicted to texting, I'll never give up my cell phone—even after I walked into that cement beam in an underground parking lot. Basically, there's no hard and fast rule when it comes to making the phone cut, so choose what feels right for your household. The good old EPA has a website listing cell phone recyclers, which include most cell phone companies: http://www.epa.gov/epawaste/partnerships/plugin/cell phone/index.htm.

I Need Some Facts to Bore my Friends With

Americans own over 234 million cell phones, and of these 70 million are reported lost every year. Where are the 70 million phones, America?

Some of them are at your local Goodwill. Next time you need a new phone, grab one from the thrift store bin. Or if a friend is upgrading, buy her old unit. You can have service transferred for little to no cost.

Recycle Everything but Your Bad Jokes

Cut to the Chase, Hippie: What's the Least I Need to Know?

More important than recycling every last scrap of plastic and aluminum (you're bound to overlook things now and then) is remembering to rinse out whatever containers you salvage before tossing them into the green bin. And don't bother recycling juice boxes or coffee cups that have a waxy coating. A lot of recycling gets sent to the landfill if it's not properly cleaned or not actually recyclable. You don't have to be obsessive about it, but be wary of contaminating your entire batch.

Intriguing . . . I Can Handle a Little More

There's a lot more to recycle than your standard plastic, glass, paper, and aluminum. When you buy products like disposable flatware or toothbrushes that are ordinarily not recyclable, look for brands that offer the service of receiving your used items for recycling. *Real Simple* publishes an extensive list of various companies that provide this service, as well as other organizations that will take your unusual recycling goods, like carpeting, film, cassette tapes, glue, sports equipment, crayons, gadgets, and even that old pair of gardening Crocs: http://www.realsimple.com/home-organizing/organizing/tips-techniques/recycle-anything-00000000006117. Not feeling motivated to recycle crayons? Me neither.

I Need Some Facts to Bore My Friends With

The United States is the number one trash-producing country in the world, with 1,609 pounds per person per year. This means that 5 percent of the world's people generate 40 percent of the world's waste. If every American recycled just one-tenth of their newspapers, we would save about 25 million trees a year. And recycling one aluminum can saves enough energy to run a TV for three hours. So just recycle your Monday night football beer can and you're even.

Upcycling vs. Downcycling

Cut to the Chase, Hippie: What's the Least I Need to Know?

Upcycling, downcycling, here's the rule: If you're kicking a milk jug to curbside collection, you're downcycling. If you're using your next-door neighbor's old garage window frames to replace the ones in your kitchen, you're upcycling.

Intriguing . . . I Can Handle a Little More

"Downcycling" refers to any process, like commercial recycling, that converts old materials into lesser versions of themselves—like the aforementioned plastic milk jug turned into a low-grade park bench. Upcycling is much more hands-on—it's when you take those same materials, say, the milk jug, and give it new, instantaneous life without degrading the quality of the material—like using it as a temporary flower pot or watering can.

As far as upcycling goes, the sky's the limit. You can rework an old dress, make broken dishes into garden art, turn used jelly jars into drinking glasses, cut up old carpet pads for tennis shoe insoles, forage through industrial construction detritus for subfloor planks for your new back deck or materials for new lawn furniture. You get the idea.

I Need Some Facts to Bore My Friends With

Upcycling extends beyond the bounds of domestic use. Many manufacturers are shaping their business strategies around its principles. One outdoor apparel company in the Pacific Northwest promotes what they call "closed loop manufacturing"—they source apparel factory rejects that would otherwise head straight for the dump. You know you've always wanted a neon orange laptop sleeve made from unsold wetsuits. Just don't buy me one.

All Plastics Are Not Created Equal

Cut to the Chase, Hippie: What's the Least I Need to Know?

Even a "safe" plastic, like the kind used for yogurt cups and food wrap, begins to leach after it's been worn down or exposed to heat. Don't reuse water bottles or old tofu paté tubs in the kitchen. Basically if you're using any plastic there is a chance that chemicals are entering your body. Just stick with wood or glass if you can.

Intriguing . . . I Can Handle a Little More

All plastics come with a price. As a result of the 2008 BPA scare, in which the government concurred that Bisphenol-A (BPA) was leaching into baby bottles and interfering with hormone levels, most manufacturers now provide us with a key to help us assess what we're up against. Ever notice those numbers wedged inside the recycling symbol on the bottom of your ketchup bottle? The digits stand for the seven varieties of plastic available on the market, and indicate which are safest and easiest to recycle. Only plastics #1, #2, and #4 are guaranteed BPA-free.

Containers marked with a "1" are those most commonly collected by curbside recycling programs and effectively reincarnated. These porous plastics are susceptible to absorbing bacteria, and should really be for one-time use.

Milk jugs, soap and cleanser containers, and cosmetics and other bathroom bounty are typically marked by a "2" and are the safer bets for your health and home as plastics go. Many children's toys also fall under this classification.

A "3" plastic might boast being microwave safe, but it's a claim worth ignoring. Polyvinyl Chloride (PVC) is as tough as rhino skin against the elements, but weakens exponentially when exposed to heat and can interfere with neurological development upon exposure. In addition, it's nearly impossible to recycle—and when it is produced or burned it creates dioxins, which are cancer causers and about as toxic as you can get. It may also cause asthma, damage to the central nervous system, liver, and kidneys, as well as cause nausea, headaches, allergies, etc. And while phthalates (the plasticizer sometimes used to soften PVC) have been banned from children's toys, sippy cups, and teethers, it is still found in flooring and some sporting goods, and has been linked to hormone disruption in children and fetuses.

Plastics "4" (flexible, like grocery bags) and "5" (durable, like utility buckets) are heat resistant and more household-friendly, which is fortunate, since a lot of these two end up in our mouths: drinking straws, squeeze bottles, and Ziploc bags are among them. While "4" is difficult to recycle, "5" (polypropylene)—which includes wide-necked containers like yogurt cups and medicine bottles—is benefiting from an increased acceptance at the curbside.

Styrofoam, or polystyrene, lucky plastic "6," is a no-brainer. Steer clear. It's toxic and just ends up in landfills.

Finally, plastic "7" includes "everything else"—mostly plastics that were invented after 1987. The use of plastic in this category is at your own risk, since we don't know what's in it.

I Need Some Facts to Bore My Friends With

Better than any of those numbered plastics, compostable plastics break down into carbon dioxide, water, and other substances, without leaving any nasty residue. When I buy disposables, I go for utensils and cups made of corn. But I stick with wood or glass when I can because using things only once is overrated.

Happy Green Holidays

Cut to the Chase, Hippie: What's the Least I Need to Know?

Before you start blasting "Jingle Bells," bear in mind that a lot of holiday décor and even presents were likely made from plastic, in a sweatshop, and will wind up in a landfill before the new year. Merry Christmas!

Intriguing . . . I Can Handle a Little More

I am actually writing this the day before Thanksgiving and just days after Butterball workers were caught on undercover video footage abusing turkeys. It's not the first time we've seen this kind of thing, but this round was especially gruesome and included one worker allegedly stomping on a turkey's head until it exploded. Even when turkeys aren't abused routinely, some farms electrocute and slit still-conscious birds' throats as a method of slaughter.

Not to be a Grinch, but the humble Christmas tree tops the list of all of that holiday cheer that'll soon wind up in the trash. And don't think a plastic Christmas tree is going to do your family any favors—sweatshop city, shipping costs from overseas, and chemical treatments render these anti-green. If you can keep plants alive, choose an indoor evergreen that grows slowly—you can decorate it for various holidays—for example, if you're really into Presidents' Day and can't wait to show off those Eisenhower ornaments. When it comes to lights, go energy-efficient, such as strings of LED lights certified by Energy Star. These use up to 75 percent less energy than regular twinkle lights. Add a timer to the mix to maximize energy-friendly cheer.

It's not just winter holidays that can be a bummer. There are hidden environmental costs lurking in things like fireworks, which are pretty to look at but are just as dangerous to nature as they are to your body. Until recently, "eco-friendly fireworks" was an oxymoron; these days, new types of pyrotechnics can be made of fewer toxic metals and give off less smoke, which is better for the air, soil, and groundwater. Still, you don't want to be downwind of most fireworks displays, and the same goes for backyard pyrotechnics.

And don't get me started on the toxic wear-them-once costumes and all of the candy at Halloween and Easter. Okay, now I know I'm just a killjoy. No one wants pennies or organic raisins when they trick-or-treat—unfortunately.

I Need Some Facts to Bore My Friends With

Antimony, used to make white fireworks, can cause damage to the lungs, heart, and stomach. Barium, for green explosions, is terrible for your internal organs. And perchlorates, oxygen-rich chemical compounds, which help fireworks burn, have been linked to thyroid problems. For the most part, these harmful side effects occur only if you're exposed repeatedly to fireworks—but it's good to know. It's a big enough concern that major regular fireworks operations, such as those connected to theme parks, are looking for healthier alternatives to the chemical-rich bursts. Monterey, California, recently switched their Fourth of July festivities from fireworks to lasers for a healthier option. May the force be with them.

clean garden

Weeds are flowers too,

once you get to know them.

—Eeyore, from A. A. Milne's *Winnie-the-Pooh*

four

I Don't Garden, but You Should

A friend of my son's has a vegetable garden, and lo and behold, those kids and mine are delighted to eat lettuce or even chard when they know it came from someone's personal garden and that their grubby little hands have picked it. The great thing about gardening is that you don't have to have a green thumb to avoid pesticides, insecticides, or bad fertilizers. This is one of those areas where you can just not buy something and suddenly you're greener—which is about my level of greenness here.

What I mean is I'm no expert in gardening. My daughter spent all year growing a thriving little plant at school and it was dead in two weeks at my house. We all have our skills.

Composting: Sifting Through It

Cut to the Chase, Hippie: What's the Least I Need to Know?

Composting isn't just for children of the sixties. Entire cities—like Boulder, Colorado; San Francisco, California; and Portland, Oregon—now offer curbside compost pickup. Whether you live on a ranch in Montana or in a closet-size apartment in New York, you can compost with ease.

Intriguing . . . I Can Handle a Little More

Not only is composting more and more common, but the contents of your compost pile can include more than you might imagine. Sure, you can throw your potato skins and banana peels into the heap, but you can also break down teabags, shredded paper, cardboard, and coffee grounds. About three-quarters of the materials we toss in the trash can usually be composted, and it takes about the same amount of time to throw a bunch of lemon peels into the compost as it takes to walk a bag of garbage to the trash bin outside.

Compost is generally used for fertilizer, replacing the expensive store-bought kind. Pour your pile of "black gold" onto your flowers and vegetables and you'll get a higher yield and at the same time protect your plants from diseases. If you live in an apartment and have no interest in planting flower boxes or nursing spider plants, you can still collect your food scraps in an enclosed composting bin and sneak the resulting fertilizer into a nearby park or find a community garden that'll gladly take it off your hands.

The most basic method of composting is to throw everything compostable into a pile in your yard, which can be contained by chicken wire or wood. It should be at least three feet by three feet.

I Need Some Facts to Bore My Friends With

Avoid putting dairy, meat, or fish products in your compost. While all of that stuff will break down over time, it'll go rancid first, and just because you're a hippie doesn't mean you have to be a stinky hippie.

Organic Gardening: Getting Dirty

Cut to the Chase, Hippie: What's the Least I Need to Know?

You don't need to have a green thumb or even a lot of open space to grow your own vegetables. If you've got a plot of about three feet by three feet, you've got enough room to get started. If not, there are plenty of community gardens that welcome new members willing to spend a couple of hours a week planting, weeding, watering, or harvesting.

Intriguing . . . I Can Handle a Little More

Seeds are dirt cheap. Organic gardening saves money, reduces your carbon footprint, and reduces the pesticides that go into your local water and air (pesticide count: exactly zero). And a zucchini that was harvested five minutes ago will definitely taste better than one cut from the ground sometime last month.

So find yourself a flat plot of good soil that's unused and free of weeds and install a raised bed. Because when beds are up off the ground, the soil in them warms up faster in the spring, so you can plant earlier. It's easier than it sounds: Just mound the soil up, or build frames from scrap materials like stones, untreated lumber, or cement blocks, and fill the area inside with soil.

Before you plant, add some compost and organic fertilizer to your plot. Go back to your local hardware store and pick up a home soil test to find out whether the dirt requires some assistance: You might need limestone or peat moss to correct acidic or alkaline soil, or some manure, bone meal, or kelp to add essential nutrients.

I Need Some Facts to Bore My Friends With

Most plants prefer to be started inside in 60 to 70 degree Fahrenheit temperatures, and transplanted outdoors in late spring.

I'm Donald Trump

Install fancy grow lights or fluorescent tubes indoors to start your seedlings in early spring. Once they sprout and develop their second set of leaves, have your gardener move your fledgling plants into larger, indoor peat pots, and keep them under the lamps for another few weeks. After that last frost, your gardener can transfer your baby plants to their raised beds outside. Your friends will just *adore* that your frisée was grown on the estate.

Okay, I've Got My Own Place, but I've Also Got Credit Card Debt

A great way to maintain your soil and keep your garden healthy year after year is crop rotation. Every plant takes a different set of nutrients from the ground, so it'll be hard for your carrots to flourish if you plant them in the same place every year.

I'm Sleeping on My Friend's Couch and Eating Ramen Noodles

Actually, organic gardening is perfect for you. Seeds cost next to nothing. You can pay your friend for her couch with fresh vegetables, and your body will thank you—ramen can be pretty taxing over the long haul.

Container Gardening: Because You Don't Need a Yard to Grow Things

Cut to the Chase, Hippie: What's the Least I Need to Know?

Whether it's a tiny flat or a penthouse suite, apartment living is no excuse to avoid growing your own cilantro. It's called "container gardening," and it's as simple and self-explanatory as it sounds.

Intriguing . . . I Can Handle a Little More

Unless you live in a closet, you can grow vegetables and herbs from the comfort of your own home (basement dwellers will likely need grow lights). All it takes is a container and a little room, which can be on the fire escape, in a window box, or on a windowsill. Start small: Try a hardy plant in a hanging pot, like licorice vine or mint. If you can keep that alive, graduate to more delicate herbs. And you'll have enough licorice vine to . . . well, nothing. Who needs licorice vine? Your containers can be anything from terra-cotta or glazed pots to an old coffee can, milk jug, wood box, or even a colander. One rule: Any container you choose needs to have holes punched in the bottom to give your plants room to drain and breathe.

I Need Some Facts to Bore My Friends With

Every plant has slightly different soil requirements, which you can find out about on the seed packet, from a local nursery, or online. But organic potting soil should work well for most of the little green guys. Or try a DIY blend: Mix two parts each of sand and peat moss to one part of compost and add a half cup of garden lime for every five gallons of soil mix.

who knew? what happened to the honey bees?

In 2006, beekeepers around the world began sounding the alarm over a disturbing trend: Bees were abandoning the hive, leaving hatchlings, honey, and even the queen behind. Dubbed "Colony Collapse Disorder" (CCD), this affliction has caused a third of U.S. bee colonies to vanish, with similar numbers in Europe. Why should we care? Because dozens of fruits and vegetables *will not grow* without pollination, and nothing's more effective than bees for doing the job. There are various theories as to the culprit, but recent studies point to neonicotinoids—insecticides related to nicotine—which mess with bees' natural homing abilities.

Fertilizers and Other Nice Names for Guano

Cut to the Chase, Hippie: What's the Least I Need to Know?

Unless you have freakishly perfect, nutrient-rich soil just lying around your backyard, you'll probably need to add some fertilizer to the mix in order to grow healthy, happy vegetables and other plants.

Intriguing . . . I Can Handle a Little More

You'll be the one eating the yields of your crops, so it's in your best interest to ensure that the carrots, zucchini, and squash in your garden are getting all of the nutrients they need. First, take stock: Buy a soil test at a local nursery and measure the levels of nutrients in your soil, which should have an equal balance of nitrogen, phosphorus, and potassium. Once you know what's missing, you can find a fertilizer that's high in the deficient mineral(s) to correct your soil.

I Need Some Facts to Bore My Friends With

It's best to use organic fertilizer, or else your lovely arugula will be tainted with toxins. Organic fertilizer treats the soil as a living organism, and helps create solid soil structure, retaining moisture and breaking down organic matter. Plus, soil treated with chemical fertilizer can become compacted and hard to work.

While store-bought organic fertilizer is fine, you can go a step further and take a homegrown approach. Depending on what your soil test tells you, there are different fixes that don't require a trip to the nursery. Powdered eggshells are rich in calcium, coffee grounds provide nitrogen, and wood ashes contain potassium.

I'm Donald Trump

Go with bat guano all the way, obviously. Exotic mail-order bat feces cost a pretty penny, but your plants will shimmer with that high-quality, high-class glow that only bat crap can provide.

Okay, I've Got My Own Place, but I've Also Got Credit Card Debt

Liquid seaweed fertilizer could be the answer to your depleted soil. While worms don't love the salt content of seaweed, liquid fertilizers are especially useful for feeding vegetables by applying it to the plants themselves, bypassing the soil altogether. This method, called "foliar feeding," is a quick shot in the arm for plants and delivers nutrients through the leaves at lightning speed.

I'm Sleeping on My Friend's Couch and Eating Ramen Noodles

Hopefully you're already composting. For an even better fertilizer, soak bags of aged compost in water and strain the resulting liquid, then add it to your garden beds. Do not drink as tea.

Pesticides: What Keeps Away Insects Poisons You, Too

Cut to the Chase, Hippie: What's the Least I Need to Know?

To use chemicals on your plants is to defeat the whole purpose of gardening at home. You might as well buy your vegetables from the local supermarket and be done with it.

Intriguing . . . I Can Handle a Little More

Herbicides and pesticides don't just off your weeds and bugs. They can travel far from your yard and contaminate areas hundreds of miles away. That's bad news for frogs and turtles, and it's bad news for us: Acute pesticide poisoning is responsible for a large number of human deaths, especially in developing nations. The herbicide glyphosate can cause eye, skin, and upper respiratory problems. Other herbicides have been linked to nausea, headaches, chest pains, fatigue, and even cancer. If that's not enough, soil fertility is compromised as well, which starts a vicious cycle: You'll need to use pesticides and herbicides to keep insects and weeds away because now your soil won't have its natural defenses.

Mercifully, you can scare off hungry insects and obnoxious weeds with natural alternatives. Try mulching, plant diversity, growing native and locally adapted plants, rotating your crops, and introducing beneficial insects to your garden. If you stay on top of it, weed-pulling by hand isn't difficult. And make sure you're not putting anything in your compost or mulch that contains weed seeds. You can also search online for natural pesticides and you'll find ways to use things like garlic, onions, and tobacco in your garden. I mean, they keep me away—why wouldn't they work on bugs?

I Need Some Facts to Bore My Friends With

Herbicides account for the highest usage of pesticides in the home and garden sector, with over 90 million pounds applied to American lawns and gardens every year. Of thirty commonly used lawn pesticides, nineteen are linked to cancer, thirteen to birth defects, twenty-one with reproductive effects, fifteen with neurotoxicity, twenty-six with liver or kidney damage, and eleven have the potential to disrupt our hormonal system. Kids are at particular risk: The National Academy of Sciences estimates 50 percent of lifetime pesticide exposure occurs during the first five years of life. Studies also find that dogs exposed to herbicide-treated lawns and gardens have increased risk of bladder cancer and double the chance of developing canine lymphoma. I know, I know, your lawn is really pretty.

who knew? dirt worshipper

Soil conservation means enriching soil responsibly and preventing erosion and pollution. Industrial runoff, wildfires, road construction, and slash-and-burn agriculture all contribute to erosion, which can ultimately lead to desertification. Definitely don't use nitrogen-based chemical fertilizers, commonly used in industrial agriculture. They leach contaminants into the water table and air, and harm the natural microbes that help soil and plants, including earthworms. Agricultural methods that deplete the amount of carbon the soil can store contribute to global warming, too. "Living soil" has the appropriate nitrogen/phosphorus/potassium balance, and is alive with microorganisms like protozoa, fungi, bacteria, worms, and insects. It's becoming harder to find, but you can make it yourself with the help of a compost pile.

Rain Catchment: I Mean, Shouldn't Water Be Free?

Cut to the Chase, Hippie: What's the Least I Need to Know?

Fresh, additive-free water is catch as catch can. It's easiest to collect H_2O for the purpose of watering your garden, but some homes rely solely on rain catchment systems to supply water for everything, from toilets to the kitchen sink.

Intriguing . . . I Can Handle a Little More

Rainwater is soft, pH neutral, and salt-free. Unlike tap water, rainwater is your garden's best friend. But an average rainfall might drop only a half inch of water into your garden. If you set up a rain barrel, that half-inch rainstorm can deposit 300 gallons of water or more into your rain bank, which can be used over the course of a couple of weeks.

Rainfall harvesting relieves some of your dependence on municipal water systems, getting you that much closer to off-the-grid, independent living. Start with a rain barrel, which should hold around 50 to 80 gallons and have a spigot for filling watering cans, as well as a connection to a hose. Barrels should be made from oak or another dark material, to prevent light penetration and bacterial growth. Top your barrel with a screen to keep debris out and prevent egg-laying mosquitoes and other insects from making your standing water their larvae base.

Ensure that your roof doesn't contain asbestos before setting your system, as you'll be collecting rainwater from the roof. If you're worried about a particularly dirty roof, empty the rainwater from the barrel the first time you collect it; water from the second round and thereafter should be fine to use for your garden.

I Need Some Facts to Bore My Friends With

A typical household can save up to 1,300 gallons of city water during dry summer months alone by collecting their own. Most rainfalls will overfill your barrel, so try installing an overflow pipe close to the top of the barrel to divert water away from your house's foundation during heavy rains. I have one of these at my house. Warning . . . .children love a big barrel of water with a spout.

Raising Chickens

Cut to the Chase, Hippie: What's the Least I Need to Know?

Rescue some unwanted chickens, let them poop in your garden to naturally fertilize your vegetables, and gather their eggs for omelets and quiches if I haven't convinced you to go vegan yet. It's that simple.

Intriguing . . . I Can Handle a Little More

Chickens are most fertile and egg-rich between ages one and four, so start out with chicks. Check local classifieds to find people who are trying to get a few off their hands. Chicks need a small space to live; this could be a cardboard box or a small animal cage. Either way, cover the bottom of their space with pine shavings and keep the temperature in their habitat between 90 and 100 degrees—these are beach babies! You can place a 100-watt bulb in a corner of the box or cage.

When chicks start growing adult feathers, it's time to build a coop and a chicken run—a cozy place where your chickens can sleep and lay their eggs and a larger space where they can run around protected from neighborhood predators. Each chicken needs at least three square feet of coop, so if you have five chicks, make your henhouse about twenty square feet or more and cover the floor with pine shavings. If you're going to let them hang outside during the day, they'll want about four or five square feet each, but they'll need a house for nighttime either way.

Chickens should be fed twice a day. You can buy their food at pet or feed stores, or toss them kitchen scraps, such as vegetables, fruit, bread, and grains. They'll also eat pesky insects and weeds and naturally fertilize your garden if you let them wander around during the day.

I Need Some Facts to Bore My Friends With

Most city ordinances ban keeping roosters, but a few hens shouldn't be a problem—they don't make a ruckus in the mornings like their male counterparts. Still, check out the rules where you live. Chef Paula Deen was recently cited for the five rescue hens she kept in a custom-built chicken coop on her property in Wilmington Island, Georgia. Her heart was in the right place. Who knew Chatham County was anti-chicken?

Goats Are the New Cow

Cut to the Chase, Hippie: What's the Least I Need to Know?

I'm not much for the domestication of animals, but if you can rescue a goat and give it a better life, go for it. Store-bought goat's milk is a little intense for many palettes, but fresh goat's milk tastes closer to cow's milk—just richer and sweeter. If you own your own goat, you'll have goat's milk daily during milking season, without the funky taste of the market-bought variety.

Intriguing . . . I Can Handle a Little More

Goats have a bad rap: They'll eat anything, including tin cans and your entire garden. The first part isn't entirely true, but the second part is, so fence off your vegetables and flowers if you don't want your new goat to gobble them up.

Goats are actually much more selective about their eating habits than you might think: They don't like to eat anything that's been lying around on the ground, and they prefer shrubbery and trees to grass. They're more "browsers" than grazers, and they'll eat the bark off your trees before they'll mow the lawn. Ideally, feed them bales of hay and alfalfa pellets and they'll let you milk them without too much complaint.

I Need Some Facts to Bore My Friends With

When you're choosing a goat, it's rescue a *pair*: Goats are social creatures, and a single goat will get lonely, stressed, and act out like a teenage boy, destroying things and trying to escape. (Even if your goat has a buddy, these guys can be rebellious—take the pygmy pair who recently tried to board a city bus in Vancouver, Washington. Three times.)

Speaking of urban goats, most cities do allow a couple of goats in the yard, but look for Lamanchas or Oberhaslis for a more peaceful goat-owning experience. And unless you live on a farm with tons of open space, don't even think about Nigerian goats—they're super noisy. These are the things your mother never taught you. Unless she was a Nigerian goat farmer.

Urban Farming: Let Someone Else Do the Digging

Cut to the Chase, Hippie: What's the Least I Need to Know?

Time is money, so if you're too busy to play around in the garden for a few hours every week and you've got some change to spare, hire someone else to do it for you. Better yet, lend your yard to an urban farmer and share the crops.

Intriguing . . . I Can Handle a Little More

Urban gardeners and sustainable landscapers can create low-impact, low-maintenance gardens that are tailored for your particular locale and climate. They'll be able to impart lots of knowledge about resource conservation, buying seeds and other garden essentials locally, and building with re-used and recycled materials.

I Need Some Facts to Bore My Friends With

More and more communities are also becoming home to urban farming cooperatives that connect home owners with folks who are willing to farm their land. The "Farm My Yard" campaign in Portland, Oregon, has been particularly successful: http://www.farmmyyard.org. Home owners sim-ply put a sign outside that says "Farm My Yard" and pretty soon a wet hippie shows up to do just that.

I'm Donald Trump

Ask your CEO friends who did their rose gardens and hire that person to start yours. If you just want vegetables, hire a star sustainable landscaper and watch your garden grow before your eyes—without the mess of dirty fingernails and scratched-up knees. Ultimately they'll save you more than they cost you—and who doesn't love a garden that pays for itself?

Okay, I've Got My Own Place, but I've Also Got Credit Card Debt

A sustainable landscaper doesn't have to cost an arm and a leg. If you're willing to share your crops, consider your soil a tradable resource. Pay your gardener—perhaps the person eating ramen and living on your couch—with vegetables.

I'm Sleeping on My Friend's Couch and Eating Ramen Noodles

Ride your bike around looking for those "Farm My Yard" signs and you'll have something to eat besides ramen. No bike? Go to: http://www.farmmyyard.org.

clean health and beauty

Joy is the best makeup.

—Anne Lamott, novelist and nonfiction author

five

Pretty Shouldn't Be Poisonous

I've been as guilty of this as anyone else: I like certain products so I try to ignore all the toxic ingredients I've never heard of. Maybe I tell myself, *Hey, how bad could it be? It's legal.* But the health and beauty products on the shelves are full of poisons that can cause everything from rashes to hair loss to cancer. It takes some effort, but I've found nontoxic alternatives to all my favorite lotions, toothpastes, and makeup brands. But read carefully—I just found that a shampoo I have with the word "organic" in the brand name doesn't have a single organic ingredient in it.

Once you've squared away your cosmetics, start thinking about your overall health. When it comes to keeping in shape, the easiest, cheapest approaches work best for me. I have a friend I walk with and we call it therapy—I show up frazzled, thinking life is crashing down around me, we walk and we talk and everything seems to come into perspective. Until the next time when life is crashing down around me again.

Lipstick Hugger: Friendly Cosmetics

Cut to the Chase, Hippie: What's the Least I Need to Know?

There are few laws limiting the chemicals cosmetic companies can use in personal care products—basically if you can use it to clean your sink it's legal to put it in your makeup. Buy from cosmetics companies that have specifically committed to healthy ingredients and eco-friendly packaging. I like Hourglass and Tarte. Other safe brands include Devita and Juice Beauty. Skin Deep, The Environmental Working Group's online cosmetics database, keeps all the most up-to-date brand information: http://www.ewg.org/skindeep.

Intriguing . . . I Can Handle a Little More

Lead in your lipstick? Carcinogens in your foundation? Yep, your makeup might be poisoning you, not to mention anyone you kiss. Products labeled "Pure," "Natural," and even "Organic" don't have to be. Look for the "USDA Organic" seal or the "Made with Certified Organic Ingredients" seal. Companies don't even have to test products—and when they do, they often rely on inhumane and even deadly animal testing. While the chemicals in any one product may not cause us obvious harm, it always makes sense to minimize our use of chemicals linked to cancer, birth defects, and other health problems. At work, this tomboy has to wear makeup every day. So I'm in a constant beauty versus health battle with my makeup artist. I'm winning for the most part (except when she sneaks in some chemical-filled blush to poison me). And she has to admit that the nontoxic vegan products work—even if the admission is under her breath.

I Need Some Facts to Bore My Friends With

Buyer beware: Maybe they didn't call that new shade of lip color "vampire" for nothing. Dangerous shades can get into your bloodstream. Recent science indicates there is no safe level of lead exposure, but according to a study by the Campaign for Safe Cosmetics, 61 percent of lipsticks contained lead, with levels up to 0.65 parts per million. Lead-contaminated brands include drugstore brands as well as high-end choices. The FDA released a follow-up study that found high lead levels in leading lipsticks. Still, the FDA hasn't taken action to protect consumers. Hmm . . . a big government agency not protecting us? How strange.

Hair Care: Green Locks Aren't Just for Punk Rockers

Cut to the Chase, Hippie: What's the Least I Need to Know?

Hair dye can cause allergic reactions, asthma, and non-Hodgkin's lymphoma to clients as well as hairstylists. Relaxers marketed primarily to African American women often contain lye or other harsh chemicals that can lead to breakage and extreme hair loss. If a product is absolutely chemical free it will have that brown and green USDA organic seal. Short of that, a good stylist can advise you on a few chemical-based brands that do less damage to both your tresses and the environment.

Intriguing . . . I Can Handle a Little More

If your shelves are full of hair products with ingredient lists that read like science experiments, replace them with products that are made with plant-based ingredients—but instead of sending old products straight to the landfill, replace them as they're used up, and recycle the old containers. Once you've gotten rid of the old products, it's time to find certified organic brands that work for you. Keep in mind that many brands are labeled "Natural" because they contain one organic ingredient along with a slew of synthetic ones.

Shampoos and conditioners with certified organic ingredients are relatively easy to find—I use JASON—but organic hair dye is often sold only in natural food stores and online. Look for mineral-based dyes, too—they're made without ammonia or synthetic preservatives so they don't contaminate the water supply when rinsed down the drain. If conventional hair dye is a must, you can stay pretty clean by avoiding ammonia, peroxide, PPDs, coal tar, lead, toluene, and resorcinol.

If you prefer the salon to the kitchen sink, you can keep it natural at a green salon that uses only certified organic products—some even practice sustainability on a larger scale, using recycled wood flooring or energy-efficient lighting.

I Need Some Facts to Bore My Friends With

Who knew you could recycle that old hair dryer? Yep. Hair straighteners, dryers, and curlers are considered "e-waste," or electronic appliance waste, and can be recycled at any e-waste event or facility. I know, an e-waste center sounds like a hell of a party, but you'll feel good about yourself later. Check out the Environmental Protection Agency's website for "eCycling" options in your area: http://www.epa.gov/epawaste/conserve/materials/ecycling/live.htm.

I'm Donald Trump

Make an appointment at a green salon every six weeks.

Okay, I've Got My Own Place, but I've Also Got Credit Card Debt

Opt for just highlights or lowlights at a conventional salon—healthier because the dye doesn't touch your scalp.

I'm Sleeping on My Friend's Couch and Eating Ramen Noodles

Skip hair dye altogether. Some pretty cool trendsetters sport gray—think Stacy London. Some even before nature gives it to them—think Kelly Osbourne. Or try my personal solution: Do nothing and pluck out that pesky gray hair that pops up every time I have a stressful day. Despite what they say, thirty more have never grown back in its place. Although yesterday my mom looked at my head and said, "Do you have gray hair?!" So maybe I'm not plucking fast enough.

Tooth Care: Chew on This

Cut to the Chase, Hippie: What's the Least I Need to Know?

Totally eco-friendly toothbrushes are labeled as such, but if you find yourself shopping for a toothbrush at a truck stop outside of Podunk, just look for one that's BPA-free.

Intriguing . . . I Can Handle a Little More

When it comes to brushing our teeth (yes, I hope you do that at least twice a day), we have to worry about what's in the brush and what's in the paste. When it comes to the paste, I use Nature's Gate. Or pick another organic brand, such as Weleda or JASON. Fluoride is probably the most controversial ingredient in toothpaste, and while many eco-friendly brands make a fluoridated version, it's an ingredient that shouldn't be taken lightly. A waste product of the phosphate fertilizer industry, flouride is a known carcinogen, a neurotoxin, and an endocrine disruptor. Modern research doesn't support the idea that it reduces risk of cavities, and with so many states adding fluoride to the water, most people just don't need it.

With toothbrushes, in addition to BPAs, watch out for the ones that contain PVC and phthalates. *How bad could it be if it's allowed in a toothbrush?* Pretty bad, actually. My kids looked at me like the Tooth Fairy had died when I handed them their wooden toothbrushes, but the chemicals in some standard brushes are downright scary. Bisphenol-A (BPA), a hormone-mimicking chemical, is linked to cancer, infertility, type 2 diabetes, and obesity. Polyvinyl chloride (PVC) is a plastic that produces numerous toxins, particularly when chewed on by young kids. And phthalates, used to improve plastic flexibility, are linked to hormone disruption, early puberty in girls, and fertility issues. And here we were just trying to keep our pearly whites gleaming.

Some PVC- and BPA-free toothbrush brands include: Oral-B, Baby Buddy, Green Sprouts, and Preserve. Preserve brush handles are made from recycled yogurt cups and even come with a pouch to mail back the used brush for recycling.

I Need Some Facts to Bore My Friends With

Fluoride isn't the only questionable ingredient in toothpaste. The Environmental Working Group tested 192 types of toothpaste and found that a third contain ingredients that may pose cancer risks, 44 percent contain harmful impurities, and almost 20 percent contain ingredients that increase our exposures to carcinogens in other personal care products. When you look at your toothpaste ingredients, you want to put it back on the shelf if it's sporting artificial colors (anything that starts with FD&C), cocamidopropyl betaine (a potentially carcinogenic foaming ingredient), parabens, sodium lauryl sulfate, or the pesticide triclosan.

Skin Deep: Your Moisturizer Just Got in My Bloodstream

Cut to the Chase, Hippie: What's the Least I Need to Know?

When checking ingredient lists, look first at the fragrance—if a product is scented only with essential oils instead of synthetics (such as diethyl phthalate, di-n-butyl phthalate, and benzyl butyl phthalate), it's likely to be less toxic. I use JASON and California Baby sunscreen and Weleda and John Masters Organics bare lotion.

Intriguing . . . I Can Handle a Little More

According to the Environmental Working Group, 56 percent of sunscreens contain the chemical oxybenzone, which is linked to hormone disruption and potential cell damage that may lead to skin cancer. Ironic, isn't it? Endocrine disruptors—chemicals also found in detergents, food, toys, and pesticides—harm the reproductive and immune function in both humans and wildlife. A 2009 U.S. Geological Survey study linked endocrine disruptors to increased numbers of mutated fish found in rivers throughout the country.

Alternative ingredients to look for that are not harmful to aquatic life and also protect well against UV rays are avobenzone, titanium dioxide, zinc oxide, and mexoryl. Aubrey Organics, Devita, and JASON all make great sunscreens using healthy ingredients.

I Need Some Facts to Bore My Friends With

If you're trying to steer clear of animal products, triple-check those ingredient lists on your skin care products. Goat placenta, anyone?

I'm Donald Trump

No problem—buy all your skin care products in Europe. More than a thousand cosmetic ingredients known to cause cancer, birth defects, and reproductive problems are banned for use in cosmetics in the European Union. Only eleven of these chemicals are off-limits in the United States.

Okay, I've Got My Own Place, but I've Also Got Credit Card Debt

If you're not paying for your undereye cream in euros, the best way to avoid harmful ingredients is to look for products that don't have too many. Keep it simple, and avoid some of the more well-known chemical offenders, like parabens, sulfates, phthalates, nitrosamines, and anything petroleum based.

I'm Sleeping on My Friend's Couch and Eating Ramen Noodles

Natural oils make great moisturizers. For drier skin, get off the couch, walk to your friend's kitchen, and use his olive or coconut oil. Hopefully he won't notice next time he goes to sauté something. For oily skin, use grapeseed, hazelnut, or jojoba oil. If you're feeling fancy, add a few drops of lavender essential oil to this simple go-to facial moisturizer. For skin toners, try witch hazel for acne or equal parts water and aloe vera juice for irritated skin.

Paint It Red: Manicures and Pedicures

Cut to the Chase, Hippie: What's the Least I Need to Know?

Some of the easiest-to-find eco-friendly nail polishes are also the most affordable. Ask for water-based and vegan-friendly options—they often go for as little as $7 a bottle in beauty supply stores and online.

Intriguing . . . I Can Handle a Little More

I'm far from perfect. I've certainly been known to get a manicure in a regular nail salon, but nail polish, polish removers, and artificial nail products contain a host of toxic chemicals that can cause cancer, reproductive harm, and occupational asthma. Nail salon workers are particularly at risk for exposure, as they work with these products all day every day, often in poorly ventilated spaces. Three chemicals of particular concern in nail polishes are toluene, formaldehyde, and dibutyl phthalate. This "toxic trio" has been phased out of most polish brands in recent years, but be sure your brand doesn't contain them.

The good news: Eco-friendly salons are popping up all over the country, and many nail polish manufacturers are beginning to reformulate their products, removing some of the toxic chemicals. I keep it simple and clean by buying natural nail polishes, which my kids love. They paint everything but their nails.

I Need Some Facts to Bore My Friends With

In 2007, *Time* magazine named nail salon work one of the worst jobs in the United States because of the toxic products used in most shops. But the industry isn't going anywhere—it rakes in about $6 billion every year. The 350,000 manicurists in the United States, the vast majority of whom are women, are exposed to a constant dose of toxins for eight or more hours a day.

A study conducted by the University of Massachusetts, Lowell, found that nail workers suffered from a host of health issues, including musculoskeletal disorders, breathing problems, headaches, and rashes.

I'm Donald Trump

Go to a green salon and get the works—and have them rub your neck while they're at it.

Okay, I've Got My Own Place, but I've Also Got Credit Card Debt

At a regular salon, ask for water-based polish or bring your own.

I'm Sleeping on My Friend's Couch and Eating Ramen Noodles

Do your nails at home with affordable vegan-friendly polish.

who knew? pretend you're a river

The Clean Water Act of 1972 attempted to regulate pollutants and keep rivers and their aquatic life-forms alive, but studies show that 39 percent of America's river system is contaminated. The law is now middle-aged and has gotten pretty gray and saggy. Besides new and old pollutants, rivers are hurt by climate change. There are little things you can do to help. For example, don't pour household chemicals on the ground or down a storm drain, don't use herbicides or pesticides in your garden, and don't flush your old medicines down the toilet—take them to a pharmacy or other facility that accepts medical waste.

Clothing: Because Even Hippies Shouldn't Have to Go Naked All the Time

Cut to the Chase, Hippie: What's the Least I Need to Know?

You can't go wrong choosing American-made organic cotton. Check the tag, it will tell you.

Intriguing . . . I Can Handle a Little More

Just because you're green doesn't mean you have to dress in burlap sacks and Birkenstocks—unless you like that look. I've certainly gone down that path and there's proof on the Internet, but no need to look.

Eco-conscious fashion has come a long way. For the couture set, annual eco-fashion weeks in New York and Vancouver showcase the latest. The Centre for Sustainable Fashion in London even offers degrees in fashion and the environment.

For the more casual fashionista, several American companies offer sweatshop-free and nontoxic clothing for the whole family, addressing the two main issues to think about when we get dressed: Where was this thing made and what's it made of?

The greenest outfit, of course, is the one you already have. Mend it, have it altered, recycle it, or donate it if you must—but don't throw it out.

If you do need something new-to-you, shop vintage, exchange, and thrift stores. When only brand new will do, look for organic cotton, hemp, soy, and bamboo. Look for locally produced clothes that are certified "Sweatshop Free" and stay away from clothing that is "Dry Clean Only" because traditional dry cleaning requires the carcinogen PERC. For the delicate items you do have, consider hand-washing. If you'd rather not do it yourself, two alternatives to dry cleaning are professional wet cleaning and liquid carbon dioxide (CO_2) cleaning.

I Need Some Facts to Bore My Friends With

Why should you care if the cotton's organic? Fully 10 percent of all the pesticides and 25 percent of herbicides used in the world are used in the production of cotton. Conventionally produced clothing is chock-full of these pesticides, toxic dyes, and dangerous chemicals, including formaldehyde.

As for leather, I'm not the perfect vegan. I'm guilty of the occasional boots or jacket, but if you're looking to take your vegan lifestyle to the next level, avoid wearing leather, fur, wool, and even silk. Silkworms, just like cows, sheep, and creatures with soft fur are raised by the millions, only to be killed in the process of harvesting a part of their body. In the case of cows raised for leather, some are skinned alive. In the case of sheep raised for wool, they're bred to have wrinkly skin, which produces more wool but also attracts flies and maggots. And the poor little silkworms are boiled alive in pursuit of that fancy new blouse. Just saying.

I'm Donald Trump

Frequent all the eco-fashion shows in London. Send me a ticket while you're at it.

Okay, I've Got My Own Place, but I've Also Got Credit Card Debt

Stick to American-made, sweatshop-free brands, like American Apparel.

I'm Sleeping on My Friend's Couch and Eating Ramen Noodles

Thrifting is cheap, hip, and earth-friendly—and everything comes back in style eventually . . . except maybe that polyester button-down with the big collar. You can get rid of that. And don't give it to anyone. You can afford to not be so green on this one.

Yoga: Just Say Aum

Cut to the Chase, Hippie: What's the Least I Need to Know?

The health benefits of yoga include increased strength and flexibility, reduced stress, and the management of all kinds of health conditions, from depression to cancer to insomnia. Buy a yoga mat that's made out of biodegradable materials instead of PVC.

Intriguing . . . I Can Handle a Little More

I was in my twenties and living in New York when I first decided to give yoga a try. I'd heard about the physical and psychological benefits. By slowing down in the middle of a hectic day, I figured I could attune to my surroundings and give myself the pause to remember why I'd chosen a sustainable lifestyle to begin with.

There are so many different styles of teaching and varying yoga traditions, it can take time to find the right studio and teacher. Some classes, including most you'll find at fitness clubs, are strictly exercise. Others are more geared toward meditation and mindfulness. So shop around. The most common types of yoga are kundalini and hatha. Kundalini is about breath and awakening energy in the spine. Hatha yoga is the yoga of physical poses known as asanas. Within hatha yoga there are many subdivisions that people practice, including Bikram yoga (which is done in a heated room), Iyengar yoga (which focuses on alignment), and power or flow yoga (which comes from ashtanga, a specific sequence with a vinyasa performed in between poses).

I say lean toward a class with at least some meditation, since it's been shown to actually change brain structure that, in turn, makes you less stressed and happier.

I Need Some Facts to Bore My Friends With

The word yoga means "union" in Sanskrit—union between consciousness and action, the emotional and physical, between movement and breath. *The Yoga Sutras of Patanjali*, an ancient Sanskrit text about the philosophy behind yoga, draws connections between the way we treat ourselves, how we relate to one another, and how we relate to the environment as a whole. Focusing on health and interconnectivity, yoga is about as eco-friendly as it gets when it comes to exercise.

To understand the influence that the environment has on the practice of yoga, notice how many asanas are named for plants, animals, and natural phenomena—"tree pose," "cobra," and "thunderbolt" are examples.

World's Oldest Yogi

She's ninety-three years old, but Tao Porchon-Lynch still hits the perfect pose every time. She's been named the world's oldest yoga teacher by Guinness World Records and has no plans to quit. "I'm going to teach yoga until I can't breathe anymore," she says, "then it's going to carry me to the next planet."

I'm Donald Trump

Hire a private yoga instructor to come to your home and develop a personal practice just for you.

Okay, I've Got My Own Place, but I've Also Got Credit Card Debt

Group classes are a great place to begin—and many studios offer community classes at half price.

I'm Sleeping on My Friend's Couch and Eating Ramen Noodles

Pick up a yoga book or video at your local library and develop a routine you can do every day—maybe with the friend whose couch you're sleeping on. Set an intention before each practice . . . say, getting off your friend's couch. Unless you're happy there—in which case, more power to ya, brother.

Tai Chi and Qigong: Not Just for the Freaks in the Park

Cut to the Chase, Hippie: What's the Least I Need to Know?

Low-impact, slow-motion, you can get started with Tai Chi or Qigong even if you've been sitting on the couch eating potato chips for years.

Intriguing . . . I Can Handle a Little More

You've seen the old people in the park doing it. But the benefits of Tai Chi and Qigong are significant for practitioners of any age or fitness level. Long described as "meditation in motion," Tai Chi and the related—and simpler—Qigong are now being touted as "*medication* in motion." Considered a healing art in China, there's growing Western scientific evidence to confirm that these mind-body practices prevent and treat a slew of health problems, including arthritis, breast cancer, heart disease, and diabetes. Health benefits of Tai Chi and Qigong include improved strength, circulation, balance, flexibility, sleep, and reduced pain and stiffness.

Tai Chi and Qigong are said to balance the body's energy meridians, and I don't know about your meridians, but mine could sure use some balance.

I Need Some Facts to Bore My Friends With

How many exercise options have their own holidays? World Tai Chi and Qigong Day is celebrated on the last Saturday of April—and the holiday has been recognized by sixty countries. Now, I better start planning how I'm going to celebrate World Qigong Day because it can really sneak up on you. And I can't bear to hear myself say, "Can you believe it's World Qigong Day again?"

The practices originated as forms of martial arts in ancient Taoist China. The names of some movements, like "Wild Horse Leaps the Ravine," or "Step Up to Seven Stars," make it ideal for the dedicated green hippie. I mean, what's "elliptical machine" got on "Embrace Tiger, Return to Mountain"?

Take a Hike: Exercising Outside

Cut to the Chase, Hippie: What's the Least I Need to Know?

Take a twenty-minute walk in your neighborhood. Tie your shoelaces first.

Intriguing . . . I Can Handle a Little More

There's no greener way to get fit than to go outside and connect with your local environment—walk or bike, swim, or even rake the leaves. There are no gym fees or fancy outfits required. And you can burn just as many calories as you might at the gym—a 135-pound woman will burn some 250 calories in an hour of raking or brisk walking. Since I'm not raking, I hike the trails in Los Angeles. Just recently I decided to take it up a notch and run down a trail. It was exhilarating! I was one with nature! And then I tripped over my shoelace and went tumbling ungracefully down the mountain. I opened my eyes to four strangers looking down on me, asking if I was all right. "I'm fine," I squeaked, trying to keep my cool while I gathered my strewn-around glasses, phone, and iPod as blood poured from my hand, knees, and ankle.

I Need Some Facts to Bore My Friends With

Regular exercise increases heart and lung function as well as strength and stamina, lowers the risk of heart attacks, stroke, non-insulin-dependent diabetes, osteoarthritis, and osteoporosis. A recent study published in the research journal *Environmental Science and Technology* found that exercising in natural environments was not only associated with physical benefits, but with mental well-being. Study participants who exercised outside reported greater feelings of revitalization, increased energy, and positive engagement—along with decreases in tension, confusion, anger, and

depression. If the sun's out, you'll get the added benefit of a dose of vitamin D. This is especially important if you are overweight—a recent study found that people who are overweight are almost twice as likely to be deficient in vitamin D.

who knew? the green gym powers itself

Ever looked around at all the sweaty people exercising their butts off at your local gym and think, *What if we could harness this energy?* Okay, I've never really thought that, but one French guy did, and the idea is catching on. If you happen to be in the Netherlands and in need of a workout, try visiting the human-powered dance club, which generates its own electricity using your dance steps.

The first green gyms were built in Hong Kong, Spain, Australia, the United Kingdom, and Portland, Oregon. These facilities operate sustainably, generating all their own power. The Green Microgym in Portland has its machines plugged directly into the power grid. The place not only generates the power it needs, it sells the excess back to the local utility company.

More into DIY than gym culture? Electricity-producing exercise equipment is available online. Making an exercise bike that can power your home is actually cheap and easy. Using a training stand, direct current generator, battery, and inverter, your bicycle can be used to create and store electricity to be used in your home. If you're not interested in storing the electricity, a simpler model, involving only a bicycle, training stand, and direct current converter can be used to power electrical devices during and immediately after your workout. Pedal, get the laundry done, give your friends another thing to laugh at you about.

Sneezing Is Funny, but Allergies Aren't

Cut to the Chase, Hippie: What's the Least I Need to Know?

Two things the green household can do when faced with the misery of allergies: Check your cleaning habits and know your natural remedies.

Intriguing . . . I Can Handle a Little More

Some allergies can be life threatening, so don't defy doctors' orders, but most pharmaceutical companies still score pretty low when it comes to ecological friendliness (some are beginning to explore "green chemistry" to reduce energy use and cut production costs). Acupuncture, aruyvedic medicine, homeopathy, and other healing arts are ecologically friendly and provide relief for many allergy sufferers. Herbalism and Chinese medicine can help, too—check out the health-food store for natural antihistamines, like quercetin. Various nutrients in foods are super immunity boosters, too, so dietary changes can help clear symptoms. Add green tea, dark leafy greens, wild-caught salmon (if you're still not vegan), avocados, extra-virgin olive oil, and salt-free raw almonds and cashews to your diet. Eliminate inflammatory foods like wheat, dairy, processed foods, white flour, and sugar.

And don't blow off a clean house. Dust mites and settled pollens are the bad guys. So whip out those eco-friendly cleaning products and go to town.

I Need Some Facts to Bore My Friends With

If you have allergies that require over-the-counter meds, bookmark the watchdog websites, including the FDA. Some allergy medication side effects are severe enough to earn "black box" warnings—just as scary as they sound.

I'm Donald Trump

Invest in a whole-house air purifier system, organic cotton mattresses and sheets, and a HEPA vacuum cleaner. If mold is an issue, find where that water is getting in and stop it, get a dehumidifier and consider remediating. Have carpets, drapes, and upholstery steam-cleaned a few times a year to disinfect and destroy pests without the use of chemicals.

Okay, I've Got My Own Place, but I've Also Got Credit Card Debt

Old-fashioned remedies include ginseng, elderflower tea, or local honey for hay fever; a tablespoon of apple cider vinegar and a tablespoon of honey added to a glass of warm water for sinus and throat infections; and rosemary, roman chamomile, tea tree, peppermint, or eucalyptus oils added to a vaporizer to ease blocked nasal passages.

I'm Sleeping on My Friend's Couch and Eating Ramen Noodles

Rather than chemical-filled nose drops, buy a neti pot and use a saline solution or distilled or purified water.

Acupuncture: Poking Fun

Cut to the Chase, Hippie: What's the Least I Need to Know?

Your mother-in-law giving you a headache? Get some needles stuck in your gallbladder meridian. Even if you can't get rid of your mother-in-law, it'll probably take care of that headache, at least until her next visit.

Intriguing . . . I Can Handle a Little More

Next time you're suffering from allergies or a minor scrape or bruise, take the opportunity to try an acupuncture treatment and see if it relieves your symptoms. This traditional medicine, which has been around for thousands of years, can be used alone or in conjunction with Western medicine. A method of balancing and building energy, or "Qi," it can be used as preventative medicine to maintain the body's natural balance, or to treat existing conditions like chronic pain, colds and flus, fibromyalgia, diabetes, migraines, poststroke syndromes, allergies, asthma, immune disorders, addiction, and more. Look, I'm going to be really honest: I tried acupuncture to get pregnant and I'm not sure it really worked, but we had a dog named Biscuit that could barely walk, having one slipped disk after the next. After two failed $3,000 surgeries, they said we could try acupuncture, and I kind of rolled my eyes . . . but Biscuit was diagnosed with a wind invasion and went from a dog that could barely walk to tearing around the house like a puppy after a couple of treatments. Wind invasion cured.

Hate needles? Don't worry. Acupuncture needles are very different from those used by your Western medical practitioner when you get a shot—they're very fine and hairlike in appearance, and most acupuncture patients actually find the experience relaxing. And with community sliding-scale acupuncture clinics popping up all over the country to serve the uninsured and underinsured, we can all afford this time-tested form of health care.

I Need Some Facts to Bore My Friends With

Acupuncture isn't just an alternative health option with a low-carbon footprint, it's an intrinsically green worldview that sees humans as microcosms reflecting the natural world.

As part of the diagnostic process, acupuncturists consider how the environment affects our health. Does a patient feel better or worse when it rains? How about in a particular season? Did symptoms ease or intensify on a recent trip?

The language of acupuncture reflects this emphasis on nature. For example, you could be diagnosed with wind invasion like Biscuit if you had a cold, a pain that traveled from joint to joint, or a facial tic reflecting the unpredictable movement of wind. In addition to selecting acupuncture points for eliminating wind, an acupuncturist might offer tips for self-care: Cover your neck with a scarf or avoid sleeping in front of a fan.

Your Herbal First Aid Kit

Cut to the Chase, Hippie: What's the Least I Need to Know?

Keep an aloe vera cactus in the house and break off a piece to treat minor cuts and burns. It's a cactus. You don't even have to water it that often.

Intriguing . . . I Can Handle a Little More

Drugs and medicines, in their purest forms, are derived from natural substances. So why not skip the chemical processing and use the original healing plants? These products can be found in oil, tincture, and salve form at your local health-food store. Here are a few to have on hand:

Lavender is virtually an all-purpose remedy. It can be used as a sedative, and it has anti-inflammatory and antiseptic properties—it is helpful for anxiety, insomnia, headaches, wounds, and burns. The essential oil can be applied directly to the skin for antimicrobial purposes and speeding the healing process.

Calendula-comfrey is a great combination for its astringent, antibacterial, antifungal, anti-inflammatory, and wound-healing properties.

Arnica does wonders to heal bruises and muscle aches. Take an arnica tincture and/or apply it directly with an arnica salve or gel. (Don't apply arnica to broken skin.)

Citronella acts as a natural insect repellant.

Jewelweed or Grindelia is a must-have for hikers. Treat poison ivy or poison oak by washing the area thoroughly and applying jewelweed, which specifically neutralizes the rhus toxin. An alternative is grindelia, which can be applied directly to a rash as well.

Liquid Echinacea extract is extremely versatile because it can be ingested or applied externally, as it is both an internal and external antibiotic. In addition, echinacea provides a temporary boost to the immune system.

Ginger capsules are a great remedy for the less pretty ailments: stomach upsets, including motion sickness, morning sickness, and gas. They are also helpful for menstrual cramps.

Goldenseal capsules are a powerful antimicrobial and protect against a variety of microorganisms that can cause traveler's diarrhea. DO NOT take during pregnancy.

Meadowsweet Tincture and Aspirin are both fast-acting, anti-inflammatory painkillers. Also avoid while pregnant.

Rescue Remedy or Five Flower Formula provides quick stress relief during anything from a divorce to a traffic jam.

I Need Some Facts to Bore My Friends With

The earliest recorded uses of healing plants are found in Babylon around 1770 BC. A couple of hundred years later, in ancient Egypt, folks believed that medicinal plants even had utility in the afterlife. And a 60,000-year-old burial site in Iraq contained evidence of eight different medicinal plants, probably intended to be taken along in the afterlife. I guess they didn't take them.

Since the 1990s, plants have been reemerging as a significant source of new pharmaceuticals. Industry scientists spend their time exploring parts of the world where plant medicine remains the dominant form of dealing with illness. Areas of South America, for example, provide a treasure trove of medicinal plants.

Want more facts to bore your friends with? Get a degree in ethnobotany. I'm bored just saying the word. Allow me to formally apologize now to all offended ethnobotanists.

Take a Pill: Vitamins and Supplements

Cut to the Chase, Hippie: What's the Least I Need to Know?

There's no need to take supplements unless you have a deficiency that can't be remedied with dietary changes.

Intriguing . . . I Can Handle a Little More

Health food stores and drugstores are full of supplements, but with a healthy diet, most of us don't need them. And taking too many vitamins and supplements can actually cause harm. The newest studies show that excess calcium intake may actually be detrimental to heart health and too large of a dose of multivitamins may have a negative rather than positive health impact.

Additionally, a report from Consumer Labs found that 30 percent of multivitamins had significantly more or less of some ingredients than the products claimed, so if you do need supplements, look for the "USP verified" symbol and generally choose natural and organic vitamins, even if they cost a little bit more. The cheapest multivitamins usually have fewer ingredients in lesser amounts, and contain synthetic versions, such as dl-alpha tocopherol, a synthetic form of vitamin E. These multivitamins also rely on additives and fillers, and are processed in ways that can destroy much of the nutritional benefits.

I Need Some Facts to Bore My Friends With

The first vitamin was discovered by Dr. William Fletcher in 1905. Dr. Fletcher studied nutrition, and he realized that polished or white rice lacked something special. People who ate brown rice remained healthy, but those eating nothing but white, polished rice came down with a disease called "beriberi." He discovered that B vitamins in brown rice ward off beriberi. The moral to the story? Eat your brown rice. Because no one wants a case of beriberi.

who knew? the amazon rain forest: queen of the weather system

Mostly located in Brazil, the Amazon is the "lungs of the earth" because it produces 20 percent of the world's oxygen. Amazonia is home to thousands of animal and tree species, spanning an area that encompasses half the planet's rain forests. It basically never stops raining in the Amazon—except when it does, like in 2005 and 2010, when the worst droughts in a century dropped river levels as much as forty feet, likely thanks to climate change. Even worse, the droughts kill trees that would normally serve as a buffer against man-made carbon emissions, potentially creating a vicious cycle of climate change begetting more climate change. Want to help? Join an international organization working to protect the trees we still have and stop Amazon deforestation.

If You Must Smoke, Don't Smoke Rat Poison

Cut to the Chase, Hippie: What's the Least I Need to Know?

Kicking the pack-a-day habit is tough, but kicking the formaldehyde habit is pretty easy. The lists of additives found in ordinary cigarettes are out-of-this-world disturbing—everything from hydrogen cyanide (think gas chambers) to arsenic (rat poison).

Intriguing . . . I Can Handle a Little More

Certified organic cigarettes, such as Natural American Spirit, contain none of these highly toxic chemicals and although they may not technically be sustainable in the sense that they are still harmful to smokers, they certainly improve the health of those who would otherwise smoke cigarettes containing toxic chemicals.

Once you've kicked the arsenic and cyanide habit, if you're still not ready to quit smoking, why not switch to smokeless, electric "green cigarettes." You get your nicotine, your friends don't have to deal with the secondhand smoke, and you look completely dorky, thus removing the "cool factor"—which, admit it, is the reason you started smoking to begin with.

If all that doesn't get you to quit, here's what I did: I set my quit date a few months out. Then what excuse did I have? That was fifteen years ago, so I guess the method was effective.

I Need Some Facts to Bore My Friends With

Just as it's possible to disguise chemical-filled food and cosmetics as eco-friendly, it is possible to create packaging that makes cigarettes appear organic when they are not. One European brand recently changed the outside of their cigarettes to a more sustainable paper, and labeled their product "Eco-Friendly" without taking out any of the additives.

Still, many cigarettes are not only additive-free, but organic. Certified organic cigarette companies often contract with small independent farms to produce the organic tobacco. And by using organic growing methods, these farms do their part to prevent pesticides from entering not just a smoker's lungs but the atmosphere and waterways.

clean community

When we try to pick out anything by itself, we find it hitched to everything else in the Universe.

—John Muir, naturalist

We're All in This Together

So this isn't exactly an environmental story about me single-handedly saving the earth. But a couple of months ago a friend of mine asked me to interview him on a public stage at a local book fair. In spite of my life as a public figure and my current position as a daily talk show host on a live television show, I have stage fright and a fear of public speaking. In order to help my friend, and because I couldn't think of a good enough reason to say no, I forced myself to say yes. I proceeded to dread the event's arrival, sweating and twisting and turning emotionally for the good part of a month. But guess what? When the day came, shockingly the world didn't end. I was glad I helped him and it really didn't end up being a big deal. Maybe partially because almost no one showed up. But that's not the point. The point is that helping out usually ends up making you feel good, even if not at first.

Now in the green world, whether I'm thrifting, riding my bike instead of driving, or lending my voice to a green campaign, the things that make me feel good are often those that are good for my friends, my family, and my community. Whether your communal contribution is big or small, eco-conscious choices mean more sustainable and tight-knit communities. Hopefully that'll make you feel less alone.

Plant a Tree

Cut to the Chase, Hippie: What's the Least I Need to Know?

Trees create shade and prevent erosion. They serve as a habitat and source of nourishment for small animals. Through photosynthesis, they suck in carbon and release oxygen, thus lowering greenhouse gases. And you can hug them.

Intriguing . . . I Can Handle a Little More

A mature tree can produce as many as 260 pounds of oxygen and remove between 35 and 800 pounds of carbon dioxide each year. And you thought air didn't weigh anything. Trees also help recycle water and enrich the soil where they are planted. Of course, fruit and nut trees provide food and habitat. Finally, a grove or orchard is a sweet place to gather and cheer up—maybe the whole world isn't a dump after all.

You may have heard of "carbon offsets," where companies attempt to pay for the sins of their large carbon footprints by underwriting something ecological—often tree planting. Okay, so a polluting company isn't authentically motivated to environmentalism if all it has to do is plant a few trees to make nice. But a tree is better than nothing.

I Need Some Facts to Bore My Friends With

Trees really do love to drink up carbon. One acre of new forest eliminates about 2.5 tons of CO_2 per year. It's been estimated that one tree can generate a quarter of a million dollars' worth of oxygen generation, water recycling, and air purification. I don't know how they figure out the cost of clean air and water, but that sounds like it must buy a lot of it.

I'm Donald Trump

Adopt an entire grove. Your investment will return in ways that you can count, and others you'll have to just feel.

Okay, I've Got My Own Place, but I've Also Got Credit Card Debt

Sign up to participate in the Billion Tree Campaign. Launched by the United Nations Environment Programme in 2006, it achieved the monumental number of 12,585,293,312 trees planted in just the first five years—and it's still going: http://www.plant-for-the-planet-billiontree campaign.org.

I'm Sleeping on My Friend's Couch and Eating Ramen Noodles

Your schedule is probably pretty flexible. Participate in a tree planting on Earth Day or any day.

---

who knew? **clean air, clean life**

Mass bird death—when hundreds of birds simultaneously drop dead—can be caused by pesticides and power lines, not to mention pollutants and oil spills. This is bad for birds, but think about what it means for people, too. The air is polluted with smog, noxious gasses, ozone, and acid rain. According to the American Lung Association, 127 million Americans live in cities with unacceptable air pollution levels, based on particulate levels and ozone. If your lungs need a break, take a vacation: The least polluted city in the United States is Santa Fe, New Mexico. So drive your VW out there and rent an old adobe.

Thrifting: Not Just for When You're Broke

Cut to the Chase, Hippie: What's the Least I Need to Know?

If you've got the shopping gene, thrifting is an easy way to nab something unique at a great price—without being part of a wasteful global economy of junk.

Intriguing . . . I Can Handle a Little More

Think about it: That super-store blouse, made in a sweatshop in some faraway country, has a carbon footprint and a Misery Index that's not chic at all. It also required wasteful packaging—not to mention that twenty-five other women on the bus are going to be wearing the same thing. The thrift-store find, on the other hand, might be one-of-a-kind or vintage (and we know how well-made vintage items are compared to what rolls off the factory conveyor belt these days). It might be designer or couture (you'd be shocked what some people give away). By buying it, you're saving it from clogging up the landfill or smoking up the air in an incinerator. More likely than not, your purchase is also supporting an important charitable cause. Green, compassionate, *and* stylish.

I Need Some Facts to Bore My Friends With

There are currently more than 25,000 resale, consignment, and not-for-profit resale shops in the United States, and that number is growing by some 7 percent a year. The boundary between thrift stores and consignment shops can be blurry, but in general a consignment store may have even higher-end merchandise because the original owner will receive a small percentage of the sale.

I'm Donald Trump

Why don't you bring some joy to someone else by donating a big bag of stuff you don't really use to a thrift shop? You'll support a worthy charity and make someone's day when they find your Louis Vuitton scarf mistakenly sorted into the $5 scrap bin.

Okay, I've Got My Own Place, but I've Also Got Credit Card Debt

Instead of buying new merchandise with your hard-earned cash, take your used clothes to a re-sale store, like Buffalo Exchange or Crossroads Trading Company. If you are in a smaller town, look for a consignment shop. Trade in your unwanted clothes and get store credit in return.

I'm Sleeping on My Friend's Couch and Eating Ramen Noodles

Find out which thrift stores have colored-tag discounts, or shop on one of those all-you-can-fit-in-a-bag-for-$5 days. Or you could ask friends for their unwanted clothes and take them in to trade at a resale shop.

Free Stores: Where Everything Is, Yes, *Free*

Cut to the Chase, Hippie: What's the Least I Need to Know?

A brainchild of the anticapitalist movement, giveaway shops have a price that can't be beat.

Intriguing . . . I Can Handle a Little More

Known as giveaway shops, swap shops, free shops, or free stores, these are places where people come to exchange goods (and sometimes services) outside of the money-based economy. The free-store idea has its roots in the old-fashioned swap meet, where you trade one thing for another, but not all free shops require you to trade anything; in some, you can just show up and take what you need. Giveaway shops help the environment as much as thrift stores do—by keeping things out of landfills—but they also serve an important political and cultural statement in the richest and most wasteful society in the world, thumbing a collective nose in the face of the big global capitalist economy. If it can be given away for free, where do you get off charging *so much* for it?

Not all free shops need to take place in an actual store. A park or other common space will do fine. During the 2004 G8 Summit, several antiglobalization groups came together to create what's become the Really, Really Free Market (RRFM) movement: http://www.reallyreallyfree.org.

RRFM aims to build community and co-create trust. One of the first RRFMs was held in Dupont Circle, in globalist-friendly Washington, D.C. There is also the Freecycle network, a nonprofit organization that spurs online groups across the country where people post things they no longer want. And how cool is it to find out there's someone in your very town who actually wants that wrought-iron gate that's been cluttering up your garage or that box of rockabilly records your uncle gave you?

I Need Some Facts to Bore My Friends With

In gift societies of Pacific Island communities and elsewhere, things of value are regularly given away between individuals without any expectation of an exchange. There's also the potlatch of the Pacific Coast Native American tradition, where the family leader throws a lavish feast for the guests. The potlatch is seen as a way of redistributing wealth and restoring a sense of equality among people. It's in this same spirit of restoring equality that free stores are making a difference— and a statement. So get on board and throw a potlach or something.

Community Gardens: Grow Food, Meet Babes

Cut to the Chase, Hippie: What's the Least I Need to Know?

Community gardens can be informal—like the folks who clean up an old abandoned lot, replacing broken glass with sprouting seeds—or they can be highly organized neighborhood entities with well-developed volunteer schedules and governing boards that determine the planning and design of the site.

Intriguing . . . I Can Handle a Little More

Not so many of us are farmers anymore. And a lot of us don't have the time to grow a vegetable garden, or even a yard to do it in. Container gardening is going to get you only so far. So lots of urban dwellers and busy folks are getting into community gardens. They're places where friends and neighbors can come together to grow fresh food as well as friendship and a sense of connection. And you can't get more "buy local" than a neighborhood plot.

Some community gardens are organic, others use conventional plants. Some have teaching elements. There are also a bunch of guerrilla gardens around, which is an activist statement where gardeners grow things on land they don't have the legal right to use. While I love the idea of rebel gardeners, more and more municipalities and towns are eager to work with growing volunteers, because community gardens bring good vibes and good food.

I Need Some Facts to Bore My Friends With

Community gardens trace their ancestry to Victory Gardens, which were popular during World War I and World War II. Americans, Canadians, Brits, and Germans alike planted vegetable, fruit, and herb gardens to help with the war effort and create a larger locally grown food supply, to allow the government to focus on the military activities. Victory gardeners also could feel a sense of pride and being involved in making their country strong. It is nice to think that today we can invest in community gardens that can accomplish similar goals.

According to the American Community Garden Association, there are now an estimated 18,000 community gardens throughout the United States and Canada. Check out their website for more information about where there is a garden near you, or how to start one: http://community garden.org.

Policy Wonk: Affecting Change in Local Government

Cut to the Chase, Hippie: What's the Least I Need to Know?

Nowhere is "think globally, act locally" more real than in your own town or city's government. Decisions made around procurement, waste management, use of common spaces, transportation, energy, recycling, and other green things that cities do can be influenced by the civic-minded green citizen.

Intriguing . . . I Can Handle a Little More

Don't be shy—from letter-writing to sitting on the city council, regular people just like you are making their voices heard and influencing important policies that help make a difference for the environment.

Start by finding out when the city council or other influential local group meets. Attend. All public meetings of this kind are required by law to be posted publicly, either in the local newspaper or at a city hall. Learn how they operate and who the movers and shakers are. They're going to debate important stuff like neighborhood development, historic preservation projects, budgets and expenditures, all sorts of school decisions, and other matters. Some of these public forums allow people to sign up and speak for a few minutes.

And if the meetings aren't mind-numbing enough, visit the council's website.

If you think speaking your mind isn't really getting you anywhere, consider joining a community group that's working toward the same things you're interested in. Service groups like Rotary or Kiwanis, or church groups and other community organizations might be active, or there may be local branches of larger national organizations that are involved in your community. The Community Environmental Legal Defense Fund (CELDF) is a leading nonprofit organization helping communities navigate laws and organize to combat ecological issues concerning water, land, air, and resource allocation: http://www.celdf.org.

The next step up might be to volunteer to be on a city commission or another appointed board. When slots come up, they are publicized. Contact the mayor to share your interest and experience. If you want to take it to the next level, consider running for city council or a school board. You'll need to circulate a petition and do the paperwork—but hey, you might become mayor!

I Need Some Facts to Bore My Friends With

Even if you aren't up for running for the office, you might want to know the top ways U.S. mayors are seeking to lower their municipalities' carbon footprint. According to the U.S. Council of Mayors, they are: LED and other efficient lighting (76 percent), low-energy building technologies (68 percent), and solar systems to generate electricity (46 percent). It's a start—and if you visit the council's website, you can find out about some good projects and best practices that mayors are implementing to improve their energy and environmental usage: http://www.usmayors.org. See? And you thought you had nothing fun to do on Saturday night.

Getting Involved in Consumer Campaigns

Cut to the Chase, Hippie: What's the Least I Need to Know?

You can get involved in a variety of ways to protect yourself and other people from unsafe products, harmful services, or even false advertising. Grab an online subscription to *Consumer Reports* and start to educate yourself.

Intriguing . . . I Can Handle a Little More

When we spend our hard-earned money, we have the right to have a few minimal expectations—like products we buy that won't hurt us or other people. We might also be concerned about the impact the stuff could have on other communities or countries, or on the environment. Consumer rights activism means bringing attention to, preventing, and sometimes punishing corporate misconduct.

It's not just about avoiding rip-offs. In some cases, corporate abuse or negligence can literally be a life-or-death issue. There are many ways to get involved, from asking to speak to the manager (not always effective) to adding your voice to thousands or even millions in an online campaign to a company, organization, or elected official (better), or participating in a boycott of a company or product that is offensive (usually quite effective, because money *always* talks).

I Need Some Facts to Bore My Friends With

One of the first consumer rights organizations was founded in 1971 by Ralph Nader. Public Citizen was originally focused on issues related to transportation, health care, and nuclear power, and more recently has taken the charge on behalf of campaign finance reform and drug and auto safety. Their model of citizen activism has inspired many other consumer campaigns from organizations like Consumer Watchdog, Alliance for Justice, Center for Science in the Public Interest, the U.S. Public Interest Research Group, Consumers Union, and many others. Social media has also become a fast way for organizations to get the word out about issues they're fighting for. If there is a topic that is particularly interesting or important to you, check online to see who is working on it or a similar issue, and sign up to receive their alerts and sign a petition.

I'm Donald Trump

Hire a lobbyist to do your green bidding for you. Read Ralph Nader's fictional *Only the Super-Rich Can Save Us* and take inspiration.

Okay, I've Got My Own Place, but I've Also Got Credit Card Debt

Volunteer for an existing consumer campaign and spread the word via social media.

I'm Sleeping on My Friend's Couch and Eating Ramen Noodles

Get off the couch and pitch your tent at a corporate or government protest for one of the green consumer causes you're fired up about.

A Broad Church: Environmentalism and Faith Communities

🛒 ☺

Cut to the Chase, Hippie: What's the Least I Need to Know?

Religious leaders in the press always seem to be denying global warming or writing human-generated environmental degradation off as God's will. But in truth, people of faith are actually at the forefront of the environmentalist movement.

Intriguing . . . I Can Handle a Little More

Some religions are based in a commitment to nature. Hindus, for example, traditionally associate dharma (virtue, righteousness, and duty) with devotion to the natural environment, and ayurvedic medicine is based on a certain natural harmony of the systems. Buddhists, with their focus on the interconnectedness of all living things, embrace ecological viewpoints in their very approach to daily life. In Islam, the Koran stresses the sacredness of the environment as part of God's revelation of truth, and highlights the beauty and importance of nature.

Christianity has been of two minds about environmental issues. The old biblical notion that God gave mankind "dominion" over the earth has long been seen as a justification for taking the earth and its resources for granted. But in a world where climate change and scarcity of resources are real, it's encouraging to see that some God-fearing folks are expressing newfound environmental stewardship. The Green Bible, an English version of the New Revised Standard Version Bible, fo-cuses on environmental issues and teachings. It was originally published by Harper Bibles in 2008 and includes essays by notable clerics and environmentalists, from Desmond Tutu to Wendell Berry. Biblical verses with environmental messages are highlighted in green to help the reader get the point.

Other Christian traditions have a more overt connection to environmentalism, such as Seventh Day Adventists, who stress the importance of simplicity and wholesome lifestyles; the Amish, who continue to shun machines and other hallmarks of even twentieth-century "progress"; and the Mormons, who have been involved in some high-profile green building projects in the United States and abroad.

As for the Jewish tradition, teachings are found in the Torah, the practice of the Sabbath, and Jewish liturgy that command humans to protect the earth, the creation of God. A popular reference comes out of the book of Ecclesiastes, where God is leading the first humans through the Garden of Eden: "Look at my works! See how beautiful they are—how excellent! For your sake I created them all. See to it that you do not spoil and destroy My world; for if you do, there will be no one else to repair it."

I Need Some Facts to Bore My Friends With

Because more than 80 percent of Americans have a religion, and close to 80 percent of that group identify themselves as Christian, it goes without saying that churches can make an enormous difference in activating us around environmental preservation.

If You Lived Here You'd Be Home by Now: Small Growth Projects

Cut to the Chase, Hippie: What's the Least I Need to Know?

Small growth, an anecdote to sprawl madness, may be our best chance of saving the earth without, you know, some mass migration to Mars.

Intriguing . . . I Can Handle a Little More

Trends in housing and community building may not change as rapidly as furniture or fashion, but thankfully in the past couple of decades, we as a society are starting to look at how we live and get around, and what is best not only for the planet, but possibly for us as humans. Retreating from McMansion/car-dependent isolated living that defined *somebody's* version of "success," urban planners are re-embracing compact, urban, and semi-urban spaces that are walkable, bicycle-friendly, and want to preserve natural spaces. These small-/smart-growth communities include neighborhood schools, stores, and live-work mixed-use developments. Their stress on alternative transportation promotes health and wellness—important for a nation that doesn't get off its ass enough.

This kind of urban and semi-urban planning is a throwback to the old village square ideal, but with careful attention to how planning and resource usage can lessen the carbon footprint. And hopefully make life less anonymous and more pleasant.

I Need Some Facts to Bore My Friends With

Hundreds of U.S. mayors have signed on to the Mayors Climate Protection Agreement, first drafted in 2005. Based on targets of the Kyoto Protocol, the agreement embraces anti-sprawl land-use policies, urban reforestation, mass-transit reinvestment, sustainable energy policies, green procurement, and public education. The aim is to reduce overall municipal greenhouse gas emissions. Cities are vying for bragging rights in the number of green buildings, energy efficiency of municipal vehicles, sustainable waste- and water-management policies, and other efforts. By spring of 2012, 1,054 mayors had signed on, representing a total population of over 88,920,962 people.

who knew? green eggs and ham isn't greener than greensburg

Climate change means natural disasters are on the rise, so take a cue from the greenest town in America: Next time your house gets whisked away by hurricane, flood, twister, or tsunami, rebuild green.

In May 2007, one of the strongest tornadoes ever recorded virtually wiped Greensburg, Kansas, off the map. In the 1.5 square-mile town, population 1,574, 11 people died and fewer than a dozen homes were left standing. It hardly seemed worth rebuilding, but in the days after the disaster, the community came together and hatched a plan: Not only would they rebuild, they'd attract new residents and businesses as they did, turning Greensburg into "the greenest city in America"—the most energy-efficient, environmentally sensitive town in the country and an oasis for eco-friendly industries. Greensburg was no hippie town. People there were more likely to wear cowboy boots and drive trucks than wear Tevas and drive hybrids. But the idea appealed to the community's conservationist, common sense tendencies.

On a per capita basis, Greensburg now represents the greatest commitment to green building anywhere in the United States, with hundreds of new, environmentally friendly homes with reinforced basement "safe rooms," model eco-homes that showcase sustainable design, and energy-efficient government and community buildings. Way to take lemons and make sustainable lemonade, Greensburg.

Change Starts at Home

Cut to the Chase, Hippie: What's the Least I Need to Know?

"Living locally" goes beyond buying local—it means you're committed to, and active in, your own community. Beyond all the important things we talk about in this book, what can you do to make a difference in your neck of the woods?

Intriguing . . . I Can Handle a Little More

Even if it's just a few friends and you working together—or just your family, or just you—I hope this book is giving you some ideas that you can begin using in your own life to make a difference to protect our fragile earth. Here are a few more ideas.

Bike racks. No point making a lot of noise about going carless if your bike will be stolen the next time you head to the grocery store to buy that nice head of locally grown lettuce. Talk with a local store, a whole commercial block, or the local chamber of commerce about installing sidewalk bike racks. It not only sends the right message about the store's commitment to earth-friendly transportation, it will increase their foot traffic, literally. Mom-and-pop shops might be more flexible, but chain stores have more money. If that's the case, write the regional or home office for help—try their marketing or corporate social responsibility departments because they're always searching for ways to look good in front of the community.

So what if you can't get to the African jungle or the Amazon rain forest. Regular people just like you and me have combated issues like land contamination, lack of walkable areas or safe playgrounds, illegal trash dumping, hazardous waste sites, and food deserts—where people have no convenient means of access to places where they can buy fresh and nutritious food. You, too, can do something big and green. Great places to post notices about starting up or connecting with an environmental project in your community include the school bulletin board, churches or synagogues, the local library, an independent bookstore, coffee shop, grocery store, or co-op.

I Need Some Facts to Bore My Friends With

The U.S. Department of Transportation says that 44 percent of the nearly 1 *billion* daily personal car trips seat only one person. Start by posting a notice at your work, or some of the places listed above. You might even end up initiating the next slugging trend—another name for a casual carpool where people wait for a ride in lines at popular pick-up locations. The first one of these ad-hoc carpool systems was started in the Washington, D.C., metro area in 1975, and the Bay Area's casual carpool transports over nine thousand people per day from the East Bay into San Francisco.

clean work and money

We make a living by what we get,

we make a life by what we give.

—Winston Churchill, politician and statesman

Be That Annoying Person at Work

Because industry is such a culprit in environmental destruction, what a great place to start making changes. From small things like getting your coworkers to switch from disposable coffee cups to reusable mugs, cutting down paper use, or affecting larger issues like emissions, you *can* make a difference.

I'm sure everyone at my work is annoyed with me, but *oh, well*. They're probably getting used to my brand of annoying.

Most of us—myself included—don't come to the green movement looking to launch an environmental career or to shut down sweatshops, but little changes we can make in our workplaces do add up. And who knows, maybe we *will* end up turning to green careers. I never thought I'd be writing a book about eco-choices, but here I am.

Buy Less: The Importance of Being Cheap

Cut to the Chase, Hippie: What's the Least I Need to Know?

Instead of focusing on all the green products we can buy and consume—that new hybrid or a fancy water-filtration system—remember that living simply is usually the greenest choice of all.

Intriguing . . . I Can Handle a Little More

Organics and products in recycled or "green" packaging may cost a little more, but often the most eco-friendly options are the cheapest ones. Going vegan—one of the single most effective things you can do to save the planet—will also save you money. Shopping at thrift stores, conserving water and electricity, reusing and repurposing products instead of throwing them away—all of these green practices protect your wallet as well as the environment.

I Need Some Facts to Bore My Friends With

Ideally, when we buy a new laptop, we're selling or gifting the old one—but the reality is that a lot of our electrical waste just ends up in the Dumpster. The same goes for those bell-bottoms that just went out of fashion again and all the plastic reindeer figurines Aunt Mabel sent for Christmas.

The more stuff we acquire and get rid of, the fuller the landfills. So instead of throwing away an ugly armchair, cover it with new fabric. Donate the jeans. Buy refurbished electronics as well as vintage and classic items you know will last a long time. Find ways to make what you already have useful again. Bike instead of drive. Figure out what you can live without—does the family really need a second car? With enough conservation, you may even find you need to work less—which in turn could mean less driving, and a less hurried, wasteful lifestyle.

As for those plastic reindeer figurines, I don't know what you're going to do with those. Maybe send them back to Aunt Mabel next year. Better yet, bike them to her. It's greener.

Environmental Careers: Save the Earth, Make Money

Cut to the Chase, Hippie: What's the Least I Need to Know?

Whether you're an experienced editor, a midlevel manager, or just looking for your first job out of high school or college, green industries need your enthusiasm and your labor. Plug in your field and experience level at Sustainablebusiness.com's "Green Dream Jobs" search engine.

Intriguing . . . I Can Handle a Little More

In 2005, the green industry generated more than 100,000 jobs and employed close to 2 million people in the United States alone. And the industry is only growing.

What are a few of the greenest c₂ there? Become a solar panel installer— up solar power in homes and offices about $40 an hour. Or consider becoming a green architect. You'll get certified in green design and build with energy-efficient materials to minimize the carbon footprint. The pay isn't bad, either, with the top 10 percent earning about $100,000 a year. More interested in the big picture? An urban planner sees that entire communities live more sustainably and develops contingency plans for waste buildup and disaster management. The average urban planner makes about $65,000 a year.

I Need Some Facts to Bore My Friends With

Forbes Magazine compiles an annual list of America's greenest companies, so even if you don't work in a particularly green industry, you can feel good about the work you're doing. The Intel brains of your computer, for example, were created with renewable power.

Green Grants: They'll Pay Me for That?

Cut to the Chase, Hippie: What's the Least I Need to Know?

Green is the new black, so whether you want to install solar panels or start an organic garden, grants are available from tons of sources, including schools, private foundations, and state and federal governments.

Intriguing . . . I Can Handle a Little More

Green building is likely the most popular project proposal for grantors, but you might just as easily wind up with a grant for making a documentary about one of these building projects, or for doing something creative to restore a habitat to its natural, unpolluted state. Really, there are no limits on projects or funding, so go for it—you might end up with a hefty sum to fund your zero-waste restaurant or your sculptures made from trash.

First, come up with a concrete plan. If you're going to score a grant, you'll have to brainstorm a functional proposal that includes cost estimates. Like anything else, it's important to know your audience: Subscribe to the Foundation Center's on-line directory to find grantors' histories, including information about who landed grants in the past and what the money was used for.

Be creative, and remember that there are literally billions of dollars out there—it's just a matter of being passionate about your project and writing a solid proposal.

Armed with your eco-friendly idea, hit the Internet to find your free money. The Database of State Incentives for Renewable Energy: http://www.dsireusa.org; the Economic Development Directory: http://www.ecodevdirectory.com; and the Foundation Center: http://foundationcenter .org are fantastic resources that can help match your idea to eco-minded philanthropists and foundations.

I Need Some Facts to Bore My Friends With

You don't need a huge community project to get a grant. If you're thinking of greening your house with wind or solar power, try the American Wind Energy Association: http://www.awea.org or the American Solar Energy Society: http://ases.org for tons of helpful hints, or go straight to your utility company or state government for the hard cash. After the fact, the federal government gives cushy tax breaks to those who've installed wind or solar systems.

I'm Donald Trump

The U.S. Green Building Council's Leadership in Energy and Environmental Design (LEED) rating system is the gold standard of green building—if your hospital or school is LEED-certified with a platinum rating, that means it's pretty much solely running on the power of the earth. So fund a hospital that wants to construct a new building with platinum LEED certification. Maybe they'll even name the building after you.

Okay, I've Got My Own Place, but I've Also Got Credit Card Debt

Say goodbye to gas and electric bills by installing solar panels on your roof, and let the government or private organizations foot part of the bill.

I'm Sleeping on My Friend's Couch and Eating Ramen Noodles

Collect all those ramen wrappers and make art from them, then find a grant organization that wants to pay you for it. Or raise money DIY-style via Kickstarter.com or other indy-arts fundraising websites. There are plenty of galleries that have group shows based on environmentally friendly art these days, too—use your ramen art slant as a way to get your foot in the door.

When Your Company Is Destroying the Planet

Cut to the Chase, Hippie: What's the Least I Need to Know?

If you work for the enemy, make a plan to change it from the inside. Start by networking with co-workers to find your eco-allies.

Intriguing . . . I Can Handle a Little More

Because most employers now know that going green is good for business, associations have been formed within many major companies to find ways to green up and move toward renewable energy.

If your company doesn't have an organization like this, consider starting a group. If there's a recycling system in place that's confusing or outdated, your team can begin by making it more user-friendly. See that electrical waste is taken to an e-waste facility. Maybe you can change the trash bags from plastic to biodegradable, or the halogen bulbs to fluorescents.

I Need Some Facts to Bore My Friends With

Just because a company gives a nod to environmental issues doesn't mean it isn't toxic. While it's working on many biofuels, the agricultural giant Archer Daniels Midland has faced hundreds of millions of dollars' worth of lawsuits for various types of pollution. *Newsweek* consistently ranks it among the least-green companies in the country.

Monsanto (which hardly claims to be environmentally friendly), the leading producer of glyphosate herbicides, also consistently makes that list. Glyphosates have been linked to cancer and birth defects. There's also Monsanto's creation of genetically modified crops and the fact that as of 2001 the company was called out by the EPA as being "potentially responsible" for over fifty Superfund cleanup sites—the federal government's list of the worst uncontrolled hazardous waste sites. So if that's your employer, get busy, you've got a lot of work to do.

At least the Allentown, Pennsylvania–based electricity company, PPL, is up-front about their negligence. As they put it on their website: "PPL currently has no formal greenhouse gas emissions reduction plan in place." Can't say they're not honest.

who knew? water over the dam

More than half the world's major rivers are dammed. While this can provide hydroelectricity and prevent floods, dams also often lead to erosion, flooded forests, destroyed wetlands, and displaced people whose land is suddenly submerged. Dams change rivers into lakes, hurting native species and favoring nonnative, invasive, even harmful ones. While these are all serious issues, the World Commission on Dams is trying to focus on helping people who are affected by dam construction—often indigenous people who live downstream from dams and suffer from a higher rate of water-borne diseases.

The Big Commute

Cut to the Chase, Hippie: What's the Least I Need to Know?

According to the National Resources Defense Council, there's so much extra space in America's 140 million cars that everyone in Western Europe could fit in our extra seats. But you don't have to pick up that Swedish hitchhiker. Organize a carpool, use public transportation, or start biking to work.

Intriguing . . . I Can Handle a Little More

Get your exercise while you're traveling to and from work and you can bail on that gym membership. Walking and biking increase heart and lung fitness as well as strength and stamina, lowering the risk of heart attacks, stroke, non-insulin-dependent diabetes, osteoarthritis, and osteoporosis. Staying out of traffic jams can't hurt our mental health either. You hardly ever hear about a bicyclist erupting in road rage.

I Need Some Facts to Bore My Friends With

The average American spends about a hundred hours commuting each year.

Long commutes and traffic don't just eat away at our time, but car commutes waste energy and contribute to air pollution and global warming. Transportation of all types accounts for more than 25 percent of the world's energy use—and cars account for nearly 80 percent of that. The resulting pollution is linked to lung cancer, respiratory and immune-system problems, urban smog, and acid rain.

If that's not bad enough, atmospheric concentrations of CO_2 have increased by 30 percent since preindustrial times, and much of that increase is directly related to the burning of fossil fuels. According to the Worldwatch Institute, CO_2 levels are now at their highest point in 160,000 years, and global temperatures are at their highest since the Middle Ages—an era that, as you may recall, brought us the Barbarian Invasions and the Bubonic Plague. The effects of this global warming include rising sea levels (yes, lower Manhattan will likely be underwater in less than a hundred years); dying coral reefs; spreading of infectious diseases; melting polar ice caps—which causes loss of polar bears' and penguins' habitats; and other extreme weather conditions. But I'm sure none of this is anything to worry about.

The Eco Office: Paper Is Overrated

Cut to the Chase, Hippie: What's the Least I Need to Know?

The average American uses some 750 pounds of paper each year—and we recycle only about half of it. But most of what we used to do on paper we can now do via email. So don't print it out if you don't have to. And if you do have to, recycle it.

Intriguing . . . I Can Handle a Little More

At my work, I noticed that many of our thick briefing packets every morning were going unread, since most of the contents had already been emailed to us, so I whined and irritated everyone until they started printing smaller packets. Almost all companies are guilty of it, and most would probably change once the waste is pointed out.

Using online applications and emailing group memos to employees cuts back on paper communications. Posting training manuals on a company website keeps organizations from having to print out booklets. Direct deposit is a great alternative to paper checks, and online bills and bill payments eliminate a lot of mail.

When printing is necessary, print double-sided on recycled paper. Fax-related paper waste can be avoided by using a fax-modem, which allows documents to be sent directly from a computer, eliminating the need for a printed hard copy.

I Need Some Facts to Bore My Friends With

Americans use about a quarter of the world's paper, but it's not just trees we're plowing through to make it. Sure, paper production accounts for about 35 percent of felled trees, but paper manufacturing is also one of the largest industrial users of water, and the third-largest user of fossil fuel worldwide. Recycling one ton of paper saves seventeen trees, seven thousand gallons of water, three cubic yards of landfill space, two barrels of oil, and enough electricity to power the average American home for six months.

Sweatshops: The Real Price Tag

Cut to the Chase, Hippie: What's the Least I Need to Know?

You're probably wearing an item of clothing right now that was produced in a sweatshop. Unfortunately, this means you're supporting terrible work conditions and ridiculously low wages—some workers make as little as 43 cents an hour sewing clothes for major American retailers. All in the name of fashion.

Intriguing . . . I Can Handle a Little More

If you want to stop supporting sweatshops, the simplest thing you can do is buy used clothes. While that might seem rather unglamorous, there are tons of incredible resources for vintage and used high-end clothing that can keep you looking sharp. That said, there are an increasing number of companies producing clothing and other goods in a responsible manner. American Apparel is an example of a company that has built its whole business model around fair labor practices, but any union label means the items were produced in a reasonable work environment. Also, when you buy from a company that treats its workers well, your odds are better that they treat the earth well too.

I Need Some Facts to Bore My Friends With

If you think that buying clothing made in the United States means fair labor, think again. Southern California is actually a huge offender when it comes to factory labor laws. While there's no worldwide blanket definition of a sweatshop, the U.S. Department of Labor cries sweatshop on any factory in violation of two or more labor laws, such as minimum wages, working hours, or child labor. Labor advocates might go beyond the letter of the law, adding that fair factories should pay a living wage, provide entirely safe working conditions, regulate fair working hours, allow for medical and maternity leave, and let workers organize themselves into unions. You might be surprised how many seemingly conscious, health-based companies discourage unions or provide zero paid sick days for full-time employees.

While some people imagine that sweatshops provide work for people in desperate need, sweatshop workers' economic situations are rarely improved by the meager wages they earn. In areas where there are few opportunities for employment, sweatshops usually just start a cycle of exploitation. I mean, try living on four dollars a day anywhere in the world. And don't forget to subtract half of that for bus fare and childcare. It's not fun.

Investing: Banking on a Sustainable Future

Cut to the Chase, Hippie: What's the Least I Need to Know?

You don't have to put your money into predatory GMO superseeds or careless oil companies to save for Junior's college. Socially responsible investors (SRIs) make money while they encourage corporate practices that promote environmental stewardship, consumer protection, human rights, and diversity.

Intriguing . . . I Can Handle a Little More

A global phenomenon and a booming market in both the United States and Europe, assets in socially screened portfolios climbed to $3.07 trillion at the start of 2010. From 2007 to 2010 alone, SRI assets increased more than 13 percent, while professionally managed assets overall increased less than 1 percent. Don't ask me to explain anything in the last two sentences, but basically, green investing can make you money.

Green investing focuses on industries like renewable energy, energy storage, or biofuels. With this approach there are three main options: Buy stock in green companies, invest in a green mutual fund, or choose green exchange-traded funds.

To research publicly traded stocks, go to Sustainability-reports.com; to find out more about eco-investing, go to http://ecoinvestorguide.com.

I Need Some Facts to Bore My Friends With

Clean Edge has been tracking the growth of cleantech markets for nearly a decade, and reports that global revenues for solar power, wind power, and biofuel companies expanded from $75.8 billion in 2007 to $115.9 billion in 2008—and they're projected to reach $325.1 billion by 2018. I don't know about you, but I want a piece of that.

who knew? nuclear power and nuclear waste

Some say nuclear energy is a clean energy source because it emits no carbons. But the nightmare of nuclear accidents and the hazard of storing nuclear waste make "no nukes" the only safe policy. Japan's 2011 earthquake and tsunami caused the failure of cooling systems at the Fukushima 1 nuclear power plant, resulting in a nuclear emergency. Over 100,000 residents had to be evacuated and the total amount of radioactive material released remains unclear. Radioactivity in rain in Hawaii, on the west coast of the United States, and even in Boston has been linked to the Japan nuclear crisis. The by-product of nuclear power production is radioactive fuel, isotopes. Since some of these can take millions of years to lose just half of their radioactivity, they can't be safely stored. Exposure to radioactivity leads to genetic mutations, cancer, and death. We haven't all caught on, though: While Germany has decided to close all its reactors by 2022, China is building twenty-five new ones, with more planned.

clean transportation and travel

We have forgotten how to be good guests, how to walk lightly on the earth as its other creatures do.

—Barbara Ward, economist and environmentalist

Getting There Is Half the Fun

Okay, so in the intro I talked about my falling-apart old Prius. And, yes, I still have it. And, yes, it has aged these 200-plus pages later. There have been so many kid snacks and beverages spilled in there, I'm sure something's growing in it by now, which I guess is a new spin on green. So the dreaded new SUV is looking more appealing. I started researching the gas-guzzling machines and found one that comes in a hybrid model and in a diesel model. I found that the diesel version gets better mileage and takes bio-diesel fuel (that's the gas made from recycled vegetable oil). So maybe I am getting a new SUV—but I found a way to make it greener.

Look, will I be guilty every once in a while of not driving the extra fifteen miles for the bio-diesel? Probably. So I thought to myself, *Can I get perfect in other areas if I'm less than perfect in this one?* Which I guess is what this book is all about—doing what we can when we can and, come on, don't I get any credit for driving the Prius around for six and a half years?

Cars: A Necessary Evil?

Cut to the Chase, Hippie: What's the Least I Need to Know?

Cars pollute from the minute they come into this world until long after they wind up in a dump—*very* slowly rotting away. But they're often a necessary evil—unless you live in a big city with tons of fabulous public transport or, you know, you prefer to stay home with the curtains drawn.

Intriguing . . . I Can Handle a Little More

I applaud my friends who actually have the wherewithal and energy to bike or walk everywhere—even when it's 20 degrees out. I bike to work a couple of times a week, but most of us need cars, too. They are part of life, but there are ways to cut back use. This could just mean a long walk to a restaurant or a quick bike ride to a friend's house every so often. If you can lessen the knee-jerk reaction to jump into the car for every mundane errand, you're already ahead of the pack.

Pollution occurs when cars are manufactured, refueled, and during disposal. But the worst happens when cars are in motion. Automobiles use fossil fuel combustion to power their engines—the top cause of pollution in the world. Exhaust emissions include carbon monoxide, nitrogen oxide, hydrocarbons, and particulates. There are also evaporative emissions, or fuel vapors, which enter the atmosphere without being burned.

I Need Some Facts to Bore My Friends With

Sure, Mexico City and Los Angeles have pretty sunsets, but those pinks and oranges are visible haze, yet another reminder of the thick emissions cloud in the air. Then again, you might be surprised which cities create the most carbon dioxide: Denver, Colorado, emits twice as much CO_2 as New York City. And then there's New Delhi, India, where twelve hundred new cars are purchased every day, which is good if you need a ride to see your guru in Bombay, but a nightmare in terms of emissions.

On the upside, European cities give off less than half the emissions per person than most places in the United States. Most Euro cities are more compact and people don't need to drive as much, since there are still plenty of local grocers, butchers, and bakeries within each neighborhood.

who knew? your carbon footprint

The size of your carbon footprint depends on how much greenhouse gas you're individually responsible for releasing into the atmosphere. Specifically, a carbon footprint is the amount of carbon dioxide (CO_2) and methane (CH4) released into the atmosphere by the things we do: eating, commuting, living in our homes. You can find a footprint calculator online—this is a great way to get an idea of your overall impact on this planet and start making changes where you can. Methane contributes to global warming twenty-three times as much as CO_2, which is one reason many people limit their beef consumption—the livestock industry produces 18 percent of greenhouse gasses . . . not that I have an agenda or anything.

Learn to Drive: Don't Floor It Like SpongeBob

Cut to the Chase, Hippie: What's the Least I Need to Know?

If you'll be stopped for longer than 30 seconds—at the ATM, say, or waiting for a friend in her driveway—turn off the car. Sure, restarting costs fuel, too, but less fuel than half a minute's worth of idling does.

Intriguing . . . I Can Handle a Little More

We've all seen the Hummer idling in the sun with the driver basking in blasted air-conditioning and three cords of wood in the back, with perhaps a pair of skis or a couple of bikes thrown on top for good measure. If that driver also has a heavy foot on the gas pedal and constantly accelerates unless he's coming to a quick stop, then he has fulfilled the trifecta of dreadful driving behavior. Perhaps this picture hits a little too close to home, but it's never too late to cut back on carbon emissions and keep money in the bank.

The smartest way to go is to purchase a very small hybrid to get you around town. If that's not going to happen, at least avoid the common driving pitfalls. For example, make sure you know where you're going before you start the engine, so you don't get lost; if you do, pull over and regroup, rather than driving aimlessly in search of your destination. On that note, shortcuts are your friends—take a minute to reassess your daily routes and see if you can challenge yourself to find a shorter or less congested one to work and/or school. I'm amazed when I drive from my girlfriend's house to work—if I drive on the main street, there's bumper-to-bumper traffic, but just one street over I can fly to work in ten minutes. Haven't the drivers on the main street ever taken a minute to think about their daily commute? I'm hoping those particular drivers don't, but everyone else around the globe will.

I Need Some Facts to Bore My Friends With

Rapid starts and stops guzzle gas, so it's better to avoid rush hour altogether if possible. And although some highways allow drivers to blaze through at 75 miles per hour, you start losing money for every 5 miles per hour you go over 60, at the tune of an additional quarter or so per gallon. That really adds up if you drive 75 for an hour or more.

Biofuels: Or How to Run Your Car on Old Restaurant Grease

Cut to the Chase, Hippie: What's the Least I Need to Know?

Gas is disgusting. It's made of fossil fuels, which are essentially decomposed plant and animal deposits that have been buried underground for millions of years. So don't let the gross-out factor put you off cars that run on the oil your local diner uses to fry potatoes and chicken strips.

Intriguing . . . I Can Handle a Little More

Biofuels are from plants. There's ethanol, made from corn or sugar, and fuel made from vegetable oil. In the case of ethanol, corn or sugar is harvested, broken down via chemical reactions, fermented, heated, then refined—and voila, ethanol. All told, ethanol produces about 90 percent less toxicity for our air than gasoline. Then again, these crops come from land that could be used for food production, and are gobbling up energy in the process. It's not a perfect picture, but it's prettier than gas.

An even prettier picture involves good ol' recycled grease. Biodiesel made from recycled, processed vegetable oil is one option; you can also convert your engine to run on unprocessed veggie oil. Biodiesel is certainly easier to find—bigger cities have a gas station or two with biodiesel options—but unprocessed vegetable oil is the planet-friendlier choice, since it requires no further processing after you get it from a restaurant.

I Need Some Facts to Bore My Friends With

Any diesel car can run on biodiesel fuel produced by a reputable manufacturer with no conversion required. You'll have to change your engine a bit to run on unprocessed grease or vegetable oil—old diesel cars are the easiest to convert. If you're a true car junkie, there are kits you can buy for a do-it-yourself approach; for the average person, you'll need to find a good mechanic who knows how to convert your engine.

A final caveat: It might be illegal to drive a car that runs on pure vegetable oil in some states. Federal and state revenue agents require that you obtain special licenses, and when running your car on vegetable oil, still pay motor fuel taxes. It's a bit ridiculous, but that's the way this gas-centric country is set up—for now.

I'm Donald Trump

Buy a lightly used luxury car (if it's brand new, all the toxicity created in production is on your shoulders) and convert it to delicately sip straight-up vegetable oil.

Okay, I've Got My Own Place, but I've Also Got Credit Card Debt

Next time you get a new-to-you car, go for a diesel engine that you can run on biodiesel, or a "flexible fuel vehicle" (FFV), which can run on up to 85 percent ethanol. Models being manufactured as a flexible fuel vehicle include the Dodge Caravan, the Ford Taurus, and the Chevy Tahoe. Toyota is planning to introduce FFV models in the United States as well.

I'm Sleeping on My Friend's Couch and Eating Ramen Noodles

Convert your old beater to "greasel" status and ask your local deli for the French fry oil. If you're lucky, they'll say yes and you'll have free fuel to get around town.

Take Your Gas and Shove It: Fuel Cells and Electric Cars

Cut to the Chase, Hippie: What's the Least I Need to Know?

Whether your car is hybrid, pure electric, or fuel cell–powered, the price at the pump will be less—or nonexistent. Plug-In America's website can catch you up to speed (pun intended) on all the latest and greatest electric-based auto innovations: http://www.pluginamerica.org.

Intriguing . . . I Can Handle a Little More

Sure, you might spend more initially, but the savings pay off: If you drive about 15,000 miles per year and get an average of 21 miles per gallon, you'll need a little over 700 gallons of gas per year. Conversely, if you have a hybrid, you'll get about 40 mpg, and need about 375 gallons of gas per year. Go completely electric and the subject of gasoline is moot.

All-electric cars have larger batteries than hybrids and can usually go between a hundred and two hundred miles before they need a charge. Trouble is, they need to charge for ten to twelve hours to run the full range. They're usually more expensive than hybrids and their limited range makes them bad candidates for, say, road-tripping. Also, the production of their batteries is a super-toxic process. And there is one more thing: Electric cars are only as clean as their power source. Most electricity in the United States comes from fossil fuels, so unless you're plugging into a solar-powered or other renewable energy source, you're still producing greenhouse gases. Now that all the negatives are out of the way, let's be real: Electric cars are awesome. They're way greener than gas-powered cars. If we all really could run our cars solely on water, wind, or solar power, smog would literally disappear and the worldwide dispute on oil would die down to a dull roar. While charging can take some time, if you have a 240-volt outlet—used to power clothing dryers—at home you could cut your charge time in half. If you don't have a sweet outlet like that available, there are about five hundred charging stations in this country—which hardly compares to our hundred thousand gas stations, but hey, five hundred is something.

I Need Some Facts to Bore My Friends With

Electric cars need little maintenance and no tune-ups. This is great for the consumer, but bad for big auto corporations: In the 1990s, General Motors infamously mass produced the EV1, a completely electric car, only to repossess and destroy them all a few years later. That said, according to a 2011 Pike Research study, annual sales of plug-in electric vehicles are predicted to reach 360,000 by 2017.

Public Transportation: Get on the Bus

Cut to the Chase, Hippie: What's the Least I Need to Know?

Households that take public transportation and live with one fewer car can save more than $9,900 per year.

Intriguing . . . I Can Handle a Little More

Communities that invest in public transit reduce U.S. carbon emissions by 37 million metric tons annually, but individuals can make a pretty big impact themselves.

By switching to public transportation, a single commuter can reduce a household's carbon emissions by 10 percent—and up to 30 percent if the plan includes getting rid of that second car. When compared to other household actions that limit CO_2, taking public transportation can be ten times greater in reducing this harmful greenhouse gas.

Americans take some 10 billion public transportation trips every year. And those of us living in areas served by public transportation save 785 million hours in travel time and 640 million gallons of fuel annually in congestion reduction alone.

I Need Some Facts to Bore My Friends With

The history of mass transit on land in the United States begins in the 1830s with the introduction of horse-drawn omnibuses—or stagecoaches modified for local service. Omnibuses were bumpy rides—not as tiring as walking, but hardly any faster. Horsecars running on iron rails provided smoother and faster travel. First introduced in New York City in 1832, horsecars spread in the 1850s, thanks to a method of laying rail flush with the pavement so as not to interfere with other traffic. By 1853, horsecars in New York alone carried about 7 million riders. No need to drive alone when you can jump on your local omnibus.

Motorcycles and Scooters: Born to Ride

Cut to the Chase, Hippie: What's the Least I Need to Know?

Sure, it's sexy, but riding a motorcycle or scooter can up your green ante, too.

Intriguing . . . I Can Handle a Little More

On average, two-wheeled vehicles get about twice as many miles per gallon of gasoline. This is because their sweet little internal-combustion engines do a better job of converting fuel into energy and, well, they're smaller. But don't get all high and mighty on your hog. There's a drawback: Extracting more energy from the fuel means producing a larger amount of smoggy emissions. Cars have better emissions equipment than motorcycles and scooters, which emit about sixteen times more carbon monoxide, hydrocarbons, and nitric oxide emissions into the air. In other words, these little guys pack quite a punch—*oops*.

For lower emissions, go for a four-stroke scooter. But for all you cool kids who aren't going to be caught dead on a scooter, any bike, even with its drawbacks, is still greener than a car.

I Need Some Facts to Bore My Friends With

There are electric and fuel-cell scooters and motorcycles out there. Electric motorcycles are almost silent and have zero-emission electric motors. If you go that route, then you can be as high and mighty as you please.

Bicycles: Put the Fun Between Your Legs

Cut to the Chase, Hippie: What's the Least I Need to Know?

It's simple: Bikes are the cleanest, greenest, meanest forms of transport around—unless you want to walk.

Intriguing . . . I Can Handle a Little More

On the whole, Americans are lousy bikers: We use bikes for less than 1 percent of all urban trips. In Italy, 5 percent of trips are on bicycle, and in the Netherlands it's 30 percent. A whopping seven out of eight Dutch people over age fifteen have a bike—in Amsterdam alone there are more than 700,000 bicycles, complete with three-level bike parking lots. Think it's annoying when you have to drive a car up to the third floor to find a parking space? Try biking the three stories, then getting down the staircase in your wooden clogs. That said, Portland, Oregon, is repping the cycling community for the rest of us, with 75 miles of off-street bike paths and lots of bike rental options. New York City has a greenway that circles almost all of Manhattan, and about 400 miles of bike lanes. San Diego features a 26-mile bike loop around the Bay, and Chicago has 12,000 public bike racks; 141 miles of marked, on-street bike lanes; and 35 miles of shared-use trails. So it's not all bad in this country—unless you live in the sticks.

I Need Some Facts to Bore My Friends With

If everyone in this country who lives within five miles of their workplace (about half of us) rode a bike to work just once a week, it'd save about 5 million tons of pollution each year. That's the equivalent of the disappearance of a million cars currently on the road. And if that's not enough fun for you, there's the World Naked Bike Ride, an international clothing-optional bike ride that takes place in cities every summer. So be careful when buying a used bike.

Walk It Off

Cut to the Chase, Hippie: What's the Least I Need to Know?

If the number of kids who walk and bike to school returned to 1969 levels, we'd save 3.2 billion vehicle miles, 1.5 million tons of CO_2, and 89,000 tons of other pollutants annually. It would be the equivalent of keeping more than 250,000 cars off the road for a year.

Intriguing . . . I Can Handle a Little More

The average vehicle speed in central London is a depressing nine miles per hour, while most people can walk about four miles in an hour.

Whether you live in London, England, or Paris, Texas, the true beauty of a place is in its architecture, cafés, winding side streets, and quiet, arched doorways. Okay, maybe not in Paris, Texas, but you'll probably see something cool.

In a car, it might not be convenient or even possible to pull over at a farmer's or flea market, whereas a self-guided walking tour might yield a stumbled-upon gem that could've gone completely overlooked otherwise.

I Need Some Facts to Bore My Friends With

The slow travel movement might not have the most exciting name, but it's a pretty cool spin on traditional tourism—it's when you take time to experience destinations and local cultures thoroughly, as opposed to the way tourism usually goes, with sightseeing trips squeezing fourteen spots into the space of an hour. And it's all about putting one foot in front of the other and walking from place to glorious place.

The slow travel movement is an offshoot of the slow food movement, which began in Italy in the 1980s as a protest against the opening of a McDonald's in Rome. There's even a website, Slowtrav.com, which will tell you all about your options, primarily in France, Italy, Spain, the British Isles, and Switzerland, though there are sections on the rest of the world as well.

who knew? the forest through the trees

Half the freshwater in the lower forty-eight states flows through forests that filter and clean it. Forests also decrease soil erosion and help prevent mudslides and floods. The forest ecosystem—trees and their undergrowth—provides us with lumber and essential products used for food and medicine. Currently, at least 121 prescription drugs sold worldwide come from plant-derived sources. But we live in a world hungry for wood, beef, and sprawl, and clear-cutting has gone wild. Forest management experts are pedaling as fast as they can to implement reforestation and promote sustainable woods, but we have to do our part by making sound choices and taking action.

Air Travel: Your Carbon Footprint on Steroids

Cut to the Chase, Hippie: What's the Least I Need to Know?

Quite literally, there's nothing worse you can do for the planet than to be a frequent flyer.

Intriguing . . . I Can Handle a Little More

One transatlantic round-trip flight contributes to global warming as much as driving a medium-size car 15,000 miles a year. But let's face it: We don't all have time to go by boat.

Some airlines are jumping on the green bandwagon and limiting their impact on the planet by improving aircraft fuel efficiency and/or using alternative fuels. They might also give pilots the straightest possible routes and avoid unnecessary detours, have engines run only when necessary, reduce the weight of onboard supplies, shorten taxiing distance from the gate to the runway, and reduce idling time. Continental Airlines, for example, has replaced most of the aircraft in its fleet with more energy-efficient planes and reduced emissions from the ground equipment at their hubs by more than 75 percent since 2000. All this greenness and they won an award from the Environmental Protection Agency in 2008 for another effort—being the first carrier to use an environmentally friendly pretreatment in their aircraft painting process.

I Need Some Facts to Bore My Friends With

Remember that air travel isn't necessarily the fastest or easiest method of travel. When you factor in the commute to the airport, traffic, the slow plod through security, layovers, and delays, an eight-hour car ride might be just as good as a plane ride, time-wise.

If you must fly, shoot for a direct flight and pack light: If two hundred passengers packed five fewer pounds each, the plane would be half a ton lighter and would guzzle less fuel as a result.

I'm Donald Trump

Buy a carbon offset for your flight, which neutralizes your footprint on the global climate by reducing an equivalent amount of carbon dioxide elsewhere on the planet. This might mean money toward a methane digester at a dairy farm in Minnesota or toward a wind farm in Romania. Al Gore says he gives money to an initiative in India that replaces dirty kerosene burners with highly efficient solar units, reducing emissions as he jets around the globe. Learn more here: http://www.carbonfund.org.

Okay, I've Got My Own Place, But I've Also Got Credit Card Debt

You can still put some money toward a carbon offset—it doesn't have to be the full price of your flight. Just research the company you're donating to to make sure you're not throwing money to the wind.

I'm Sleeping on My Friend's Couch and Eating Ramen Noodles

The cheapest options are so often the greenest. If you don't own a car, more power to you. No plans to hop a plane? You're doing the planet a favor by living simply. You're good. Keep, like, dreading your hair and making all-natural deodorant and bask in the good vibes of being kind to Mother Earth.

Geotourism: Seeing the World Without Wrecking It

Cut to the Chase, Hippie: What's the Least I Need to Know?

When on vacation, instead of leaving your trash behind and insulting the people you meet by making fun of their accents, make a conscious effort to leave the place *better off* than when you arrived.

Intriguing . . . I Can Handle a Little More

According to some experts, the U.S. tourism industry is now damaging the environment at a faster clip than the combined efforts of miners, manufacturers, ranchers, and loggers. Hotels and roads are built for visitors, we generate pollution *getting to* the Grand Canyon or Disneyland, we carry all those disposable travel items. Then there's the actual damage we do when we actually arrive in paradise. Coral reefs, for example, have been devastated by tourism. Pollution levels rise. And cultural tourism encourages a kind of "staged authenticity" (Real Indians!) that anthropologists say is harmful to many communities.

There are plenty of eco-conscious tour companies that will help you see the Himalayas or the rivers of Oregon without making a mess, but you don't always need a guide on how to treat the places you visit with respect. When camping, bring gloves and pick up some of the trash left by other visitors. When selecting hotels and motels, call ahead and ask about energy efficiency—ask if they give the option to reuse your sheets and towels for your entire stay (unless you're staying for a month—gross); most do these days. And don't forget to think about the local economy. Support mom-and-pop shops over those familiar international chains and you'll have a more authentic experience while supporting the things that made your destination unique to begin with.

I Need Some Facts to Bore My Friends With

Increasing tourism is threatening the survival of ancient Egyptian ruins, destroying the natural biodiversity of the Galapagos Islands because tourists bring with them new species that compete with the natural wildlife, and even endangering tiny monkeys in the Philippines who get so stressed out by human noise and camera flashes that they're literally driven to suicide. And I thought I hated having my picture taken. It's enough to inspire you to stay home.

I'm Donald Trump

Go all out with a luxury geotour—find the right tour company on Sustainable Travel International's website: http://www.sustainabletravelinterna tional.org.

Okay, I've Got My Own Place, but I've Also Got Credit Card Debt

Plan your own trip with the help of one of National Geographic's Geotourism map guides: http://travel.nationalgeographic.com/travel/sustain able/about_geotourism.html.

I'm Sleeping on My Friend's Couch and Eating Ramen Noodles

Head for the nearest beach or forest, camp out—the friend whose pad you're crashing at probably wants a break, anyway—tread lightly, and pack up all the garbage you can find.

clean parenting

In every deliberation, we must consider the impact on the seventh generation . . . even if it requires having skin as thick as the bark of a pine.

—Great Law of the Iroquois

I Know It's Corny to Say "The Children Are the Future," but, Hey, They Are

For me, clean parenting is the most important part of this book. Here's where I get really militant and start to alienate people with my ideas. So here I go . . . Why would we have things like household cleaners and pesticides that cause everything from rashes to breathing problems to cancer when we could just as easily skip them? I mean, are the ants outside really that bad? Or are the diseases and groundwater contamination worse? I know I'm on my soapbox now, but when it comes to our children, I can't help but think we should advocate for them, fight for them. Because there will come a day when they will look at us and say, "Why did you do this?" How will we respond? "Sorry, but the hippie stuff doesn't clean as well as the chemicals do?" I don't think so. Changing what we expose our children to could literally add years to their lives. And isn't that part of our job as parents? Consider this: More than 80,000 chemicals are in use in the United States and while accounts vary, only around 400 to 1,000 have been tested for safety. And I'm not just talking about the obvious culprits—I mean chemicals in your kid's french fries.

People like to say kids are resilient, but I think that's first-class denial. As an example, girls are going through puberty *five* years earlier than they were just a century ago, probably thanks to hormone disruptors and other environmental contaminants. Boys aren't exempt, as similar findings are being reported for them too. And that's just the start. We are surrounded by chemicals that have been linked to cancer and they are showing up in our kids' blood, urine, and even umbilical cords. The truth is that most Americans grow up and deal with all sorts of diseases they shouldn't have to. It is within our power to change that.

Seven Generations: Think About Your Kids' Kids

Cut to the Chase, Hippie: What's the Least I Need to Know?

Even though you probably won't get to *meet* your great-great-great-great-great-grandchildren, you don't want to be remembered as the one who screwed everything up, do you? The Seven Generations concept simply asks that each of us consider the impact of our actions and consumption seven generations down the road.

Intriguing . . . I Can Handle a Little More

Before you decide you really can't be bothered about what kind of world people will inhabit two hundred years from now, consider how messed up things might be today if not for some of the early ecological thinkers. Joseph Priestly—no, not the important *Jason* Priestly from *90210*—was one of the first to demonstrate that plants convert carbon dioxide to oxygen, in 1774. John James Audubon's groundbreaking *The Birds of America*, published in 1828, is still considered one of the greatest illustrated books of all time. And you may have heard of a guy named Henry David Thoreau—his 1854 book *Walden Pond, or Life in the Woods*, remains one of the bestselling ecological manifestos ever, not to mention required reading for tons of tenth-graders.

I Need Some Facts to Bore My Friends With

The Seventh Generation is referred to in the Great Binding Law of the Great Iroquois Confederacy, the peace treaty that united the Five Nations of the Mohawk, Oneida, Onondaga, Cayuga, and Seneca people (and later the Tuscarora of North Carolina) into one nation, the Iroquois. The treaty is believed to have been drafted between 1090 and 1150 CE, and it had a major impact on the Founding Fathers. In 1988, the U.S. Congress passed a resolution recognizing the Council's influence on the Constitution and the Bill of Rights. But imagine if those Founding Fathers had really taken the concept to heart? What if Thomas Jefferson had written the rights of future generations into the U.S. Bill of Rights ten generations ago? Would we still be dependent on nonrenewable fossil fuels that cause global warming? What if there had been a mechanism to test the ecological value of the automobile a hundred years ago? We might not be driving these planet-destroyers around. Seriously.

Pregnancy: Your Belly Is the Perfect Eco Apartment

Cut to the Chase, Hippie: What's the Least I Need to Know?

As amazing as the placenta is, some toxins can permeate the wall or enter the bloodstream through our skin. A study commissioned by the Environmental Working Group found more than 200 chemicals in the umbilical cord blood of every newborn they tested. So do all you can to limit your exposure by adopting the greenest and most natural eating, home care, and personal hygiene regimens possible.

Intriguing . . . I Can Handle a Little More

A good first step is to rid your home of as many harsh, toxic chemical products as you can when you're pregnant. You don't even want to be handling this stuff, so enlist a friend or partner to get rid of pesticides, paints, conventional household cleaners, and anything vinyl (that crazy vinyl smell is warning enough). Have someone else change the cat litter. Ban the Teflon pans. And put that dry cleaning to rest. Avoid recreational drugs and alcohol. Duh. And it's time to go as natural as you can stand when it comes to looking good. The chemicals in perms, relaxers, and dyes easily permeate the skin and enter the bloodstream. Ditto for the creams, lotions, makeup, and other products you put on your body. There's plenty of time to look good after your kid is born. Actually, there's not. Good luck, you're the one who wanted a kid.

I Need Some Facts to Bore My Friends With

In 2004, researchers at the University of California at San Francisco examined data from the Centers for Disease Control and found that 99 to 100 percent of pregnant women had some level of polychlorinated biphenyls, organochlorine pesticides, PFCs, phenols, PBDEs, phthalates, polycyclic aromatic hydrocarbons, and perchlorates in their system.

I'm Donald Trump

Give your home an organic makeover with a full-house water purification system, lead and radon testing (look in your local area for someone you trust), and a battery of earth-friendly cleaning products. Eat all-organic food and work with an integral medicine practitioner who can help you make smart decisions about how to handle colds, allergies, and other mild ailments.

Okay, I've Got My Own Place, but I've Also Got Credit Card Debt

Eat an organic and plant-based diet as much as you can. Steer clear of big fish (higher mercury count), such as swordfish, shark, and king mackerel. Limit albacore tuna to once a week. And get rid of that pesticide you've been spraying and start making your own vinegar-based cleansers.

I'm Sleeping on My Friend's Couch and Eating Ramen Noodles

Stop eating that ramen off Styrofoam plates and washing your plastic spoons in the dishwasher. Plastics leech chemicals. If your food has to be canned or frozen, opt for frozen—too many cans are still lined with the very toxic Bisphenol-A (BPA).

Adoption: Because There Are Plenty of People Already Here

Cut to the Chase, Hippie: What's the Least I Need to Know?

Eighty percent of adoptive parents say they chose to adopt because of trouble conceiving, but plenty of eco-conscious parents concerned about overpopulation say adoption was their first choice.

Intriguing . . . I Can Handle a Little More

There are generally two types of adoption: open, where the biological and adoptive parents agree to a degree of shared information and perhaps contact; and closed, where there's no disclosure to the child, and often the adoptive parents, about the identity or whereabouts of the bio parents.

Whether opting for open or closed, the domestic or international route, some important considerations include the adoptive process itself, how much it will cost (from nearly nothing to tens of thousands of dollars), how much time it can take (in the case of transnational orphanage adoptions, sometimes years), ever-changing international laws around adoption, and any potential hot spots that can be triggered by family "lifestyle" issues, like sexual orientation, marital status, or religion.

My situation is a little different from standard adoption because my son was my son before any court told me he was. While I did adopt him legally, it was my girlfriend who gave birth to him (she also cross-adopted our daughter, who I gave birth to). Since we were already a family and I had been on the journey since his conception, it's not like he was a child I had no connection to prior to the adoption, but for those hesitating—wondering if they can love a kid when they don't have any

DNA swimming through that kid's veins—I'll tell you that when you hold the baby, it won't matter where it came from. You'll know it's yours. The love is the same. But then, having grown up with adopted siblings, I already knew that.

I Need Some Facts to Bore My Friends With

While it's not everyday parlance to hear adoptive parents say their choice was motivated by ecological decisions, there are increasing numbers of parents, particularly in the Western countries, for whom adoption is proactive action in an overpopulated and resource-inequitable world. For each human we don't add, we lighten the impact in virtually every ecological area. Rather than add one, five, or fifteen more humans to the 7 billion our planet already has to sustain, or spend boatloads of money and anguish seeking infertility treatments, their path involves providing a better life to a child who might otherwise be relegated to an orphanage or the foster-care roller coaster. So even if adoption can't save the earth on its own, it can help a kid and build a family.

who knew? population growth

Overpopulation leads to elevated levels of crime, poverty, illness, and malnutrition. We already know that fossil fuels are finite, yet we keep on partying like it's 1999. By the way, that year was a *billion fewer* people ago. At the rate the human population is growing, some experts predict we'd need a whole new world in the not-so-distant future to provide food, resources, and space for us all. So before you decide to become the next octo mom, you might want to consider not only your own finite resources, but also the earth's. Unless you already have your reservations booked on Earth II.

The Clean Nursery: Might As Well Start Right

Cut to the Chase, Hippie: What's the Least I Need to Know?

The greenest decision you can make about outfitting a nursery is to outfit it as minimally as possible. For minimal ecological impact, think baby and planet: What do babies spend their time doing? Sleeping, lying down, and putting things into their mouth. Oh yeah, and pooping.

Intriguing . . . I Can Handle a Little More

Conventional cotton—in bedding, mattresses, and clothing—contains a cocktail of pesticides, including the carcinogens cyanazine, naled, propargite, dicofol, and trifluralin, not to mention harsh chemical dyes. These can cause anything from rashes on up. Opt for organic cotton, hemp, wool, bamboo, and other sustainable fibers. Check out options at Greencradle.com.

As for toys, the safest and greenest are made of unfinished solid wood. Any finish on the wood should be natural, like beeswax, linseed oil, or walnut oil. Many stuffed animals and soft toys are made with synthetic materials treated with chemicals to make them stain resistant and fire retardant. Choose stuffed animals and soft teethers made from all natural, untreated fibers—preferably certified organic cotton with nontoxic dyes. For cool eco-friendly toys check out Thelittleseed.com.

I Need Some Facts to Bore My Friends With

It's hard to do babyhood completely plastic free, so get acquainted with which are the least nefarious (check those numbers and the little triangles on the bottom of plastic products). For example, 1 (PET plastic), 2 (HDPE plastic), and 4 (LDPE plastic), are less bad than 3 (PVC), which is linked to lung, liver, kidney, and reproductive damage and whose plasticizers make them all soft and bendy, but also link them to hormonal disruption. And don't even think about 7 plastics, which often contain BPAs, the compounds that are increasingly linked to early onset puberty, obesity, and other ills. As convenient and user-friendly as plastic is, just keep in mind that *any* plastic may disrupt your baby's fragile hormone development. Bummer, I know.

I'm Donald Trump

Invest in organic and sustainably produced baby gear, focusing on clothes and the places where babies sleep and lie the most: mattresses, changing pads, and the crib. Opt for unbleached cotton, formaldehyde-free, and ecological dyes. Buy European toys, which have higher standards regarding chemicals and safety. If you are set on plastic, check out American brands Sprig and Green Toys. The latter uses recycled milk jugs as the primary ingredient in their toys.

Okay, I've Got My Own Place, but I've Also Got Credit Card Debt

When it comes to mass-produced, plasticy products like car seats and high chairs, leave them in the garage or shed as long as possible before the birth to cut down on the off-gassing exposure. Don't get your kiddo any polyester clothes, which are made from petroleum. If you're buying used furniture, make sure the model has not ever been part of a recall and has been tested for lead safety if it's of a certain vintage. If you can't afford a pricey organic mattress, opt for a mattress cover that zips around and prevents off-gassing.

I'm Sleeping on My Friend's Couch and Eating Ramen Noodles

Keep it simple: Old towels make fine changing pads. Used baby clothes are easy to find. And consider foregoing the crib altogether by choosing co-sleeping, although you'll have to get off your friend's couch to co-sleep safely. Wear your baby in an organic cotton sling—which you can put on your registry.

Midwives Help People Out

Cut to the Chase, Hippie: What's the Least I Need to Know?

Midwives have been going about the birthing business for as long as we know, caring for both mother and baby through the pregnancy, attending to the actual birth, and providing educational and emotional support to the new mom. With a midwife, you're more likely to end up having a natural birth.

Intriguing . . . I Can Handle a Little More

There are now two basic types of midwives: certified nurse midwives, who have completed nurse training and typically work with an OB, and direct entry midwives, whose status is attained through study, apprenticeship, or a midwifery school. This latter category typically attends births at home or in freestanding birthing centers.

Both are professionals who undergo serious training, testing, and apprenticeship before they can catch a new human. As women who often have gone through birth themselves, they are also highly attuned to the emotional, psychological, and physical signposts that take place during labor and delivery. If your aim is to avoid drugs and interventions, midwives are generally more patient and willing to let nature take its course. They're also trained to identify when a birth is not progressing "normally," warranting transfer to a hospital or obstetrical intervention. During my own childbirth experience, it was helpful to have just one more person to scream at.

I Need Some Facts to Bore My Friends With

Only about one in two hundred American women now opt for home birth, but studies show that for low-risk mothers, home births are at least as safe as hospital births, and result in fewer unneccesary interventions. Midwifery was the go-to means for childbirth until the dominantly male field of medical science began pushing back against what was seen as a "primitive" or "unscientific" practice. This trend took hold with the invention of forceps in the 1600s and chloroform as a form of anesthesia in Victorian England. By the 1970s, more women in the United States, Canada, and Western nations began advocating for the return of more natural, women-centered options in childbirth.

Childbirth Isn't for Sissies

Cut to the Chase, Hippie: What's the Least I Need to Know?

Birth can be treated as a life event rather than a medical one. Low-tech labor and birth generate less waste and require less electricity—and pain medications not only can pass through the placenta to the baby but will eventually end up in our waste water system. Bet you never thought about contaminating water systems with the drugs in your pee—learn something new every day.

Intriguing . . . I Can Handle a Little More

Make a birth plan—your stated thoughts about your preferences during labor and birth—but accept that plans might change.

You don't have to allow your newborn to be immediately vaccinated when he or she is born, you can request vitamin K drops instead of a shot, and don't let them give your baby supplemental formula or glucose water while you're waiting for your milk to come in—that'll inhibit your own production and the baby can wait. You have colostrum in the meantime—the system was designed to work without interference.

When it comes to labor-pain management, the general anesthesia of yesteryear, and the "twilight sleep" of the 1950s (semisedation, courtesy of morphine), have given way to spinal pain-blockers, such as epidurals. Epidurals are no cakewalk, though, as they require an IV and a catheter, and limit our ability to move around the room. Other potential side effects range from the mother being numbed to the point where she can't push (often necessitating forceps or vacuum delivery), to a possible chemical disruption of the speed and efficacy of labor, drop in blood pressure, itching, nausea, urinary dysfunction, spinal headache, increased drowsiness in both mother and child, and more. Various types of spinal blocks become necessary, of course, in the case of a Caesarean-section birth, but for uncomplicated births, many women do try to limit or avoid pain medication.

I thought I'd go the drug-free hypnosis-mellow route, but after about a million hours of heinous labor pains and ending up howling in the backyard on my hands and knees (not exaggerating), I took the meds. I wish I could try again to go drug free, but I guess the goal is to get that baby out, after all, and sometimes even the best-laid plans need to be abandoned.

I Need Some Facts to Bore My Friends With

Anesthesia used in childbirth is usually lidocaine or bupivacaine, but in the 1800s some doctors used cocaine (giving the term "crack baby" a whole new meaning).

I'm Donald Trump

Consider using a hypnotherapy specialist, aromatherapist, acupuncturist, massage therapist, or other holistic practitioner to support you before and during labor. Hire a doula to assist you during the birth and advocate for sticking to your birth plan as much as possible.

Okay, I've Got My Own Place, but I've Also Got Credit Card Debt

Investigate how nonmedicated pain management methods—like Lamaze, Bradley, or LeBoyer Gentle Birth—can help. Pick up a birthing hypnosis CD online and listen to it throughout your third trimester.

I'm Sleeping on My Friend's Couch and Eating Ramen Noodles

Hone your skills in deep breathing, muscle relaxation, meditation, and visualization to help you weather the storm. Keep in mind that home births are cheaper than hospital births.

Postpartum Care: You Deserve a Break

Cut to the Chase, Hippie: What's the Least I Need to Know?

Thought it was all over when the baby was born? Welcome to the joys of the "fourth trimester." Exhaustion and the hormonal roller coaster can take their toll. Pamper yourself as you take care of that baby. And remember, there are herbal remedies for everything from depression to mastitis.

Intriguing . . . I Can Handle a Little More

Instead of a fancy nipple cream, try using coconut oil. It's safe for the baby and its antibacterial and antifungal properties may keep infections away. Nelsons Hemorrhoid Cream is a homeopathic alternative to the toxic stuff (ah, yes, the glamour of it). For a good postpartum tea that's organic, healing, and safe for the nursing baby, try Earth Mama Angel Baby's "Monthly Comfort Tea." Or make your own with herbs and spices, like cinnamon, red raspberry leaf, nettle, lemon balm, lavender, oat straw, and alfalfa. Other herbs that are traditionally used to support healthy milk production include anise, blessed thistle, fennel, fenugreek, and milk thistle. Speaking of milk production, I had mastitis five or six times—over and over. The first time I had that raging fever I took antibiotics, then I learned to use a battery of herbal remedies and compresses to get over it and became a pro. Between feedings, try compresses soaked in rosemary, fenugreek, and dandelion infusions. Two or three raw garlic cloves a day can ward off that fever, too—not to mention unwanted visitors. The other thing that helped was nursing in every possible position to get the clogged milk out. Gross, I know, but sometimes green isn't glamorous.

I Need Some Facts to Bore My Friends With

Gossip magazines love to show celebrities rockin' a slammin' body weeks after giving birth, but for most of us, the postpartum period lasts up to nine months. If you're breastfeeding and eating well, most of your pregnancy weight may go away within six weeks, but not always. As you're discouraged from resuming a hard exercise routine for at least two months (by the way, who's exercising before two months?), pamper yourself while you bond with your baby. If tough emotions feel overpowering, consult the Edinburgh Postnatal Depression Scale (EPDS), a ten-item questionnaire easily found online that can help women and families determine when the "baby blues" might require real support.

I'm Donald Trump

Pick up a few organic cotton sleep bras. Set yourself up with a postpartum doula, who will make sure you're well fed, hydrated, comfortable, educated, and emotionally supported while you adjust to your new life.

Okay, I've Got My Own Place, but I've Also Got Credit Card Debt

Herbal treatments include soothing aloe vera, comfrey for the perineal area, fennel to promote breast-milk production, tonic red raspberry leaf, vitamin-rich nettle leaf—try one of the several delicious tea blends for new mothers. Arnica helps with bruising and soreness, and many Chinese herbs have relaxing properties. Eat plenty of omega-3- and protein-rich foods.

I'm Sleeping on My Friend's Couch and Eating Ramen Noodles

Fill a dollar-store ketchup bottle with witch hazel dilution to irrigate your beat-up perineum, and douse maxi pads with witch hazel and pop them into the freezer overnight for the same purpose. Cabbage leaves are a great relief for breast engorgement. Frozen towels and water-filled latex gloves make cheap ice packs. Drink plenty of water, and do your Kegels.

Breastfeeding: They Drink Until They Pass Out

Cut to the Chase, Hippie: What's the Least I Need to Know?

The perfect food for your baby is the ultimate renewable resource, always ready on tap. If you can breastfeed, do it. If it's not an option, remember that plenty of snuggling can replicate some of the benefits linked to breastfeeding. In spite of the stigma associated with prolonged breastfeeding in this country, the World Health Organization recommends breastfeeding up to age two or beyond, and the average age of weaning around the world is four years old.

Intriguing . . . I Can Handle a Little More

A lot of deep science goes into the attempted mimicry of breast milk, like one recent scary experiment in China involving infusing cow's milk with human genes. There are also soy-based formulas, as well as concoctions that contain no milk at all, but cow's milk is the most common type of formula—and the production of cow's milk is responsible for 20 percent of greenhouse gas emissions globally. Add to that the issues of land usage, deforestation, soil erosion, fertilizer use, groundwater and river contamination, hormone and antibiotic overuse in bovine farming, the energy needed for the manufacturing, packaging, and distribution, all the paper and toxic dyes involved in mostly nonrecyclable packaging, and the storage in tin cans, which are often lined with Bisphenol-A (BPA). Well, suffice it to say that can of formula sure has one massive ecological footprint. Also, cow's milk is designed for baby cows that are supposed to grow big with small brains. No wonder newborns naturally turn their head and open mouth toward the boob.

I Need Some Facts to Bore My Friends With

Breastfeeding protects your baby from a long list of illnesses and allergies. Stomach viruses, lower respiratory illnesses, ear infections, and meningitis occur less often in breastfed babies and are less severe when they do happen. A recent British study demonstrated that the longer a child was breastfed, the higher his or her score in picture- and pattern-recognition, pattern construction, and naming vocabulary. The researchers suggest that breastfed kids are one to six months ahead, cognitively speaking, of their bottle-fed mates. The outcomes were particularly striking for premature babies. The researchers point to a possible benefit of the essential fatty acids found in mother's milk, the lack of the hormones and growth agents that are in infant formula, or the simple fact that breastfed babies may be cuddled more. Those exquisite snuggles are good for mom, too—studies have linked nursing to reducing breast and ovarian cancer, a lower postmenopausal osteoporosis rate, and maternal weight loss. There's also the release of oxytocin in the mother, triggered by the baby's suckling (milk ejection reflex), a natural high that won't get anyone in trouble with the cops.

I'm Donald Trump

Hire a doula to assist with proper latch-on and other nursing issues.

Okay, I've Got My Own Place, but I've Also Got Credit Card Debt

If you have any problems, engage the services of a lactation consultant. When mine showed up, I was at my wit's end and it was as if an angel cloaked in light floated through the front door.

I'm Sleeping on My Friend's Couch and Eating Ramen Noodles

Hook up with a local chapter of La Leche League International—they've been providing support and advice to moms since 1956. Like breast milk itself, their support is free.

Pump It Up

Cut to the Chase, Hippie: What's the Least I Need to Know?

Pumping and storing milk isn't quite as environmentally pristine as breastfeeding on demand, but it's a great option for those of us returning to work or spending time out of town.

Intriguing . . . I Can Handle a Little More

When we've got to be away from the baby for extended periods, prepumping breast milk is the way to go. Unless you're expressing by hand (which is great, but nowhere as fast or efficient as a motorized pump), you'll have to accept the fact that electricity is required to make that little mechanical boob-sucker do its thing. Some models work with rechargeable batteries, so that's somewhat of an improvement.

Once you've pumped, store your breast milk in glass bottles or canning jars in the refrigerator or freezer. If you do end up having to use plastic bottles for either storage or feeding, make sure they're BPA-free. Never warm frozen or cold milk in the microwave, but submerge the bottle in a bowl of hot water for about five minutes.

I Need Some Facts to Bore My Friends With

Thanks to the 2010 Affordable Health Care Act, all employers of fifty people or more must provide reasonable break time for an employee to pump breast milk for her child up to the child's first birthday. The employer must also provide a place, other than a bathroom, for the employee to pump. Some new moms are even able to convince their employers to invest in pumps equipped with personalized kits—so several mothers can share a high-quality pump. Sounds like humane factory farming to me.

Bottle-Feeding: Because We Can't All Be Milk Machines

Cut to the Chase, Hippie: What's the Least I Need to Know?

Crazy as it sounds, human milk is best for human babies. So if you can't or don't want to supply it yourself, use a breast milk bank if possible. When it comes to formula-feeding, pay attention to both the "greenness" of the formula and the chemical hazards in the bottles and nipples. Choose soy-based, organic non-GMO brands.

Intriguing . . . I Can Handle a Little More

Whether you're having difficulty breastfeeding, aren't your baby's bio mom, or you've got other circumstances that necessitate bottle-feeding, there are a few toxic pitfalls to avoid.

First, read those labels—even USDA organic formulas may contain corn syrup or rice syrup.

As for the bottles, nipples, and other paraphernalia, let's return to Plastics 101. You have to sterilize bottles after every feeding to avoid bacterial accumulation, but given that heat increases the probability of chemical compounds leaching into the milk, you want to avoid plastic bottles containing Bisphenol-A (BPA) or polyvinyl chloride (PVC), which are linked to health problems from asthma to liver and kidney damage. And while phthalates (the plasticizer or softener) sometimes used in PVC have been banned from bottles and teethers, some companies are just substituting other dangerous or potentially dangerous chemical plasticizers. Regardless of what is added, PVC is toxic to us and probably to infants and children in particular. Look for recycling numbers 1 (polyethylene terephthalate), 2 (high-density polyethylene), or 5 (polypropylene) on the bottom of the bottle if you absolutely must use plastic. Better yet, use tempered-glass bottles just like Great-grandma used to.

I Need Some Facts to Bore My Friends With

Most formulas are fortified with synthetic DHA (docosahexaenoic acid) and ARA (arachidonic acid), essential fatty acids naturally found in breast milk. Since babies can formulate these EFAs themselves, their necessity in simulated form is unproven. Further, the fatty acids are synthetically manufactured from fungi in a process requiring hexane, a neurotoxic solvent. Synthetic DHA/ARA causes gastric distress in some babies. Though consumer demand is beginning to drive alternatives in the marketplace, it's hard to find a brand that is entirely DHA/ARA-free. Check the labels; some brands have less of this "supplement" than others.

I'm Donald Trump

Hire a wet nurse or buy breast milk from a bank. Think that's weird? Remember that when you buy cow's milk–based formula, you're still using someone else's breast milk—it just happens to be a cow's.

Okay, I've Got My Own Place, but I've Also Got Credit Card Debt

Use a faucet filter to purify your water for preparing formula. I don't recommend milk-based formula, but it costs less than soy.

I'm Sleeping on My Friend's Couch and Eating Ramen Noodles

If you have insurance, see if your policy will pay for formula. If not, apply for the federally funded Women, Infants, and Children (WIC) program as well as the food stamp program and feed your baby nontoxic formula for free.

Vaccines: A Sticking Point

Cut to the Chase, Hippie: What's the Least I Need to Know?

If you're concerned about the overall safety of vaccines, discuss it with your health provider, and get a second opinion—perhaps balance the viewpoint of a Western M.D. with that of a holistic or integral practitioner.

Intriguing . . . I Can Handle a Little More

Few topics are more likely to start a fight at the playground than whether or not to vaccinate. Some parents refuse to vaccinate, citing everything from religious beliefs to libertarian concerns about government control. Others distrust drug companies and are wary about the unproven, long-term health effects of the chemicals and toxins such as mercury, which is in flu shots and emergency room tetanus shots, and aluminum which is in most vaccines. Some are convinced there's a link between vaccines and the rise in gastrointestinal ailments at best, and autism at worst. Some will vaccinate, but insist that the doses not be bundled into "cocktails," which they feel is too much for a small body. For example the MMR and the "penta" and "hepta" vaccines put five, six, or seven vaccines together with about a dozen different antigens in the same syringe. Others argue there are just too many vaccines out there.

On the other side, some people consider vaccinating to be a civic obligation, citing how vaccines have eradicated diseases like smallpox, and have almost wiped out polio. If, they argue, we don't vaccinate each generation, these diseases could make a comeback—and there have certainly been resurgences of whooping cough, measles, and mumps. That said, recent increases in pertussis is owed to waning vaccine immunity in spite of steady vaccination rates. Some people who have recent or current family ties to parts of the world where people are still dying from preventable diseases think we Americans are ridiculously fussy when we think twice about vaccinating.

I find a reasonable approach is to weigh the severity and prevalence of an illness with the possible risk of the vaccine. For example, the benefits of receiving chicken pox naturally provides lifelong immunity and is better than this ill-thought-out vaccine, which almost every European country has decided *not* to give. However, the risk of the disease becomes greater than the risk of the vaccine as a person gets older. For women, chicken pox during pregnancy can cause severe and even lethal birth defects. Vaccines for things like diarrhea I would skip altogether—unless you live in Zimbabwe. The vaccine against rotaviral diarrhea could save hundreds of thousands of lives in the third world, but is worth very little to American children.

At the very least it's worth reconsidering the way we vaccinate. Babies are born with genetic predispositions to certain ailments like diabetes, autism, ADHD, and arthritis, but it is clear that something in their environment triggers the illness. This might be pesticides, plastics, other chemicals, or, in a small group of children, possibly immunizations. Until we know more about how the immune system works we should probably stop combining five, six, or even seven vaccines in one shot and giving them to potentially vulnerable eight-week-old babies.

I Need Some Facts to Bore My Friends With

The manufacture of vaccines involves much chemical tinkering. Scientists and researchers at Texas A&M are looking into ways to build a "greener vaccine" from plant material.

Regardless of whether or not your kids get vaccinated, investigate natural immune boosters and avoid foods that cause inflammation and weaken immunity. Immunity sappers include partially hydrogenated fats, fried foods, sugar, and corn syrup. Boosters include whole grains, organic foods, and supplements like vitamins A and C, and echinacea.

Diapers: Be the Change

Cut to the Chase, Hippie: What's the Least I Need to Know?

The average kid will use as many as 8,000 diapers before she's potty trained—that's a lot of plastic. Or a lot of water. And so goes the great debate over cloth versus disposable.

Intriguing . . . I Can Handle a Little More

Consider the impact of 27.4 billion disposable diapers per year—some 3.4 million tons plopping into the landfills. That's 82,000 tons of plastic and 250,000 trees' worth of wood pulp. Other eco-concerns about disposables include pollution and waste associated with production, and the super-absorbent gels made from sodium polyacrylate that are linked to toxic shock and allergies. The EPA has issued warnings about dioxin, a by-product of chlorine used in the bleaching process, that's been linked to damage of the central nervous system, kidneys, and liver; suppression of the immune system; dizziness; rashes; respiratory problems; and maybe even cancer. If you do use disposable diapers, opt for an environmentally friendly brand like Seventh Generation.

But before we get too smug about cloth diapers, it's worth noting that cotton is one of the most wasteful crops out there, requiring enormous amounts of water, 10 percent of the world's pesticide use, and 25 percent of the world's herbicides. Washing all those cloth diapers at home uses up to twenty-five gallons of water a day. Diaper services use less water (because of their high-efficiency industrial process), but pick up and delivery impact pollution.

I Need Some Facts to Bore My Friends With

Eighty percent of American families today use disposables, a product that didn't even exist until the early 1960s.

I'm Donald Trump

Invest in the fanciest and greenest diaper service in your area. While you're at it, since you're so rich you can hire someone to change all those diapers for you.

Okay, I've Got My Own Place, but I've Also Got Credit Card Debt

Use diapers and covers made with organic cotton, bamboo, or hemp, and an eco-friendly laundry soap. Invest in a front-loading washer, which uses less water. For disposables, seek out brands that are chlorine-free, fragrance-free, latex-free, compostable, and made from sustainable materials, like corn. There are also hybrid diapers whose inner layer can be flushed down the toilet. Use baking soda instead of fragrances to cut down on the smells in the diaper pail.

I'm Sleeping on My Friend's Couch and Eating Ramen Noodles

Find used cloth diapers through an online community or recycling group. And don't worry: Most babies are toilet-trained by around age three.

Attachment Parenting in the Real World

Cut to the Chase, Hippie: What's the Least I Need to Know?

While this may not seem like a green issue right off the bat, when you have your baby strapped to you in an organic sling, nursing until she weans herself, you'll realize it is. Attachment parenting (AP) is as green as it gets from start to finish.

Intriguing . . . I Can Handle a Little More

It starts with tears. Attachment parents don't see crying as a nuisance, but as baby's only way of communicating. They stress that many "instinctive parenting" practices, like breastfeeding, making homemade baby food, sleeping with the baby, and "wearing" the baby in a sling or other carrier, mimic the way women have raised their children all over the world since the beginning of time.

But plenty of new parents find this baby-first lifestyle too intense, and frankly unrealistic.

The truth is there are a million ways to raise healthy kids, so incorporate the aspects of attachment parenting that make sense for your family, and don't let the "my way is best" parents at the playground get you down.

I Need Some Facts to Bore My Friends With

The term "attachment parenting" was coined by pediatrician William Sears, who based the practice on attachment theory. Derived from developmental psychology, biology, and other sciences, this theory maintains that infants attach to caregivers who demonstrate responsiveness, sensitivity, and close physical and emotional proximity, especially during the first six months to two years of life. This sense of security in a developing human will set the mold for how they navigate future personal and interpersonal relationships.

I'm Donald Trump

You'll find lots of earth-friendly options for your AP lifestyle, from sustainable-fabric and chlorine-free baby slings and nursing pillows to organic foods, homeopathic medicines, and more.

Okay, I've Got My Own Place, but I've Also Got Credit Card Debt

Join an AP-oriented moms' group to learn how to make your own slings and baby food.

I'm Sleeping on My Friend's Couch and Eating Ramen Noodles

I have a friend who brought the baby home with just diapers and a couple of onesies. Talk about a green way to save a lot of money and landfill space and create a bond with your baby that lasts a lifetime. For all the bells and whistles, what the baby wants and needs most of all isn't a product or gadget, it's you.

Talking to Kids About Saving the Planet

Cut to the Chase, Hippie: What's the Least I Need to Know?

Focus on fostering your kids' connection to their community and environment and not so much on scary concerns about the future. I'm not always a model mom on this one—I've talked to my kids about the toxins in toys made in China and now they ask where their toys are manfactured. So do as I say, not as I do.

Intriguing . . . I Can Handle a Little More

Kids are natural-born environmentalists, happiest when playing with bugs or leaves, stomping in the mud, or running after squirrels and birds in the park. It also helps that, like, 90 percent of kids' books are about animals doing something cute. So it's pretty easy to instill a sense of eco-stewardship in your kids.

Start where you live—explore the critters and plants in your yard or in the park. Grab teachable moments out of everyday things, like recycling bins, or picking up garbage in the street. Remind kids that little things such as turning off lights and appliances, not running the water the whole time they brush their teeth, and using rechargeable batteries really do make a difference. As kids get older, incorporate an environmental focus into those annual school science fair projects.

Global warming is one of those important issues—like sex, drugs, and peer pressure—that you'll likely want to put your stamp on instead of letting your kids hear all about it on the street. There's no need to lie and say it's not real, but focus on all the positive things we can do to keep those ice caps frozen.

I Need Some Facts to Bore My Friends With

The National Park Service operates Junior Ranger programs around the country, as do many state and regional parks. The kids learn a lot of cool facts, and it's a pretty darn proud moment when they put on the fake ranger hat and take the pledge to protect the park.

I'm Donald Trump

Commit to eco-tourism adventures that will help you build memories as a family while doing something good for the planet. Buy a membership in a national or international environmental or ecological advocacy organization instead of a new toy or game. Sound un-fun? Kids feel proud when they're doing good. Still sound un-fun? Well . . . come to me for green, go to someone else for fun.

Okay, I've Got My Own Place, but I've Also Got Credit Card Debt

Plant a vegetable garden. If all you have is a windowsill, use a container. Few acts are more environmentally powerful for kids than actually eating food they grow.

I'm Sleeping on My Friend's Couch and Eating Ramen Noodles

Borrow picture- and storybooks about nature and the environment from the library, and make a big deal about how cool and ecological it is that communities can share books.

The Vegetarian Child: Trade You a Tofu Dog for Your Lentil Loaf

Cut to the Chase, Hippie: What's the Least I Need to Know?

With their innate love of animals, many kids naturally don't wanna eat them. And despite what some people may tell you, veggie kids can grow at least as healthily as their carnivorous pals.

Intriguing . . . I Can Handle a Little More

Whether our kids are partial vegetarians (red meat–free), pescatarians (okay with seafood), lacto-ovos, or straight-up vegans, parents of meat-free kids worry primarily about three nutrients: protein, calcium, and iron. First off, Americans generally eat way more protein than we need; the CDC recommends only 13 grams daily for toddlers (that's just two servings of soy milk), 19 to 32 grams a day for older kids, and about 50 grams a day for teenagers. Second, plants provide plenty of protein. Legumes, nuts, seeds, and grains pack even more. There's also quinoa, a so-called "super-seed" that's fun to eat, acting like a grain and a vegetable at the same time. Mix a little into your rice if you're not ready for a full serving of the stuff.

I Need Some Facts to Bore My Friends With

Plant-based iron is easy to come by, too—organic tofu, lentils, black beans, tomato paste, blackstrap molasses, dried apricots, chard, and pumpkin seeds are full of the stuff. Ditto for calcium, the bone-building mineral you'll find in broccoli, sweet potatoes, dark leafy greens, great northern and navy beans, and fortified milk and juice. And take a tip from smart eaters all over the world:

Food combos like rice and beans, tortillas and pintos, and lentils and chapatti are protein-rich, cheap, and kid-friendly.

I'm Donald Trump

For your next vacation, consider a vegetarian-friendly destination like the United Kingdom, Mediterranean countries, Thailand, and of course India, where some 40 percent of the population skips meat.

Okay, I've Got My Own Place, but I've Also Got Credit Card Debt

Join a community-supported agriculture (CSA) program—a local organic farmer will send you all the seasonal, nutrient-dense fruits and vegetables you can stand.

I'm Sleeping on My Friend's Couch and Eating Ramen Noodles

Pick your own vegetables: Check out local farms that let you harvest with your kids.

who knew? penguins and polar bears

Snowcaps are melting, and the earth's polar regions feel the strain of climate change first and worst. Pollution, global warming, tourism, and a thinning ozone layer threaten penguins. In the Falkland (Malvinas) Islands, the penguin population has dropped from 1.5 million in 1932 to 420,000 today. Erratic behavior and increased die-off of the polar bears is so worrisome that even the Bush administration took action, listing the polar bear under the Endangered Species Act in 2008. The U.S. Geological Survey predicts two-thirds of the world's polar bears will disappear by 2050.

Cooking with Kids: Because They Won't Eat It When I Make It

❤️ 💵 👫

Cut to the Chase, Hippie: What's the Least I Need to Know?

Start by designating one night each week as "family cooking night."

Intriguing . . . I Can Handle a Little More

I can't cook. I mean, at all. But I'm raising my kids as vegetarians, and since we're not on the chicken nugget train, I've had to come up with edible alternatives despite my lack of culinary talent. One thing I've discovered: It's almost impossible to mess up a vegetable stir-fry. Just chop vegetables into bite-size pieces, throw them into a wok or skillet with a little olive oil, add one or two cloves of minced garlic and soy sauce to taste, and sauté for a few minutes.

Think your kids won't eat it? If they're part of the shopping and the cooking, I've found they're a lot more open to giving the final product a try. My kids help by choosing the vegetables at the market and washing them before I do the chopping and slicing, although not well. And even I can make perfectly steamed brown rice to go along with the dish—provided I use a rice cooker.

I Need Some More Facts to Bore My Friends With

Sure, including kids in cooking meals can require patience and some extra clean-up time, but experts think it is well worth the effort.

Cooking with your kids can help get them interested in trying healthy foods they might normally turn their noses up at, and gives them a feeling of accomplishment and the real sense that they're contributing to the family. Cooking might even help keep kids grounded as they grow up. Involving kids in the kitchen is a big stepping-stone to getting them to appreciate family meals. And studies show that teens who eat with their families at least five times a week get better grades in school and are less likely to have substance abuse problems.

Play Equipment: Swing on This

Cut to the Chase, Hippie: What's the Least I Need to Know?

Children should be free to swing, slide, and monkey-bar to their hearts' content, but the installation of backyard jungle gyms and other equipment can be full of toxins and hard on both children and the earth.

Intriguing . . . I Can Handle a Little More

Look for play structures made from Forest Stewardship Council–certified wood, recycled steel, and postconsumer HDPE plastic materials, made from things such as milk jugs, soda bottles, plastic cartons, plastic bags, and the like. Rubber tiles, commonly fixed underneath swings and other places likely to have falling children, can be made from old tires or even sneakers. There's a whole industry for sustainable playgrounds, from manufacturers to builders to installers. It's possible to purchase already cut, drilled, and nontoxic-stained sets made of types of wood such as redwood or Chinese cedar. Some companies will even plant trees for every playground set sold, to offset the use of wood and make their product even more sustainable. The best backyard playgrounds will also incorporate a system to catch rainwater, which can then be collected and stored for gardens.

I Need Some Facts to Bore My Friends With

Before 2003, most play equipment made of wood was treated with chromium copper arsenate (CCA), a preservative. This contains arsenic, which then wound up in the soil and, obviously, in the children. While many public playgrounds still contain CCA, and yearly repainting mostly covers and contains the toxic chemical, new and recycled materials made and installed after 2003 are much safer.

Barack Obama's daughters, Sasha and Malia, played safely at the White House on a structure made from redwood certified by the Sustainable Forestry Initiative. For every tree harvested for their playground, five trees were planted by the manufacturing company. The wood is water-sealed and includes a swing set and a tree fort. Take that, plastic playgrounds of yore. If you don't have Obama's resources, the cheapest and easiest way to build an eco- and child-friendly playground is with a recycled tire on a rope and a sandbox made with reclaimed timber and filled with environmentally kind sand.

Skip the Petting Zoo . . . Unless You're a Fan of E. Coli

Cut to the Chase, Hippie: What's the Least I Need to Know?

Petting zoos, blech—they're cruel, often unlicensed, and literally crawling with E. coli and other infectious diseases.

Intriguing . . . I Can Handle a Little More

I can almost hear you groaning now . . . *Come on, petting zoos and animals at the state fair are as down-home as it gets.* Maybe, but like regular zoos, circuses, farm attractions, and all traveling animal displays, petting zoos mean animals as *entertainment.* Even if that ostrich isn't confined in a cage the same way the elephants are at the zoo, there's still something creepy about paying to see an animal strut its stuff in an unnatural, captive environment. And while many states require petting zoos be licensed, regulation standards are all over the map. Some of the pathogens regularly present at petting zoos include E. coli, salmonella bacteria, hepatitis, shigella, listeria, fungi, and parasites. The elderly, pregnant women, people with compromised immune systems, and children are most at risk for infection.

If you do visit a petting zoo or a farm, stress hand-washing.

I Need Some Facts to Bore My Friends With

A 2003 study by the University of Nebraska tested livestock at twenty-nine county and three large state fairs, and found E. coli in more than 13 percent of beef cattle, 5 percent of dairy cattle and sheep, and 3 percent of pigs and goats. The research lead, Jim Keen, was inspired to study E. coli bacteria disease because his own niece endured a near-fatal case of bacterial contamination from visiting a petting zoo. If you think the circus is a good alternative, think again. Animal abuse is rampant—no government agency monitors training sessions. Ringling Bros. and Barnum & Bailey circus has failed to meet even minimal federal standards for the care of the animals. In 2011, the U.S. Department of Agriculture slapped the parent company of the "Greatest Show on Earth" with a record penalty for alleged animal rights violations. Want to see for yourself? YouTube "Ringling Beats Animals" to see trainers do just that, thanks to PETA's undercover investigation. Still think it's the "Greatest Show on Earth"?

I'm Donald Trump

Hook up with a reputed conservation organization to explore wildlife ecology travel opportunities.

Okay, I've Got My Own Place, but I've Also Got Credit Card Debt

Satisfy your kid's craving for petting by adopting a rescue dog or cat.

I'm Sleeping on My Friend's Couch and Eating Ramen Noodles

Borrow animal books or DVDs from the library or watch the nature shows on TV.

Birthdays: Towing the Party Line

Cut to the Chase, Hippie: What's the Least I Need to Know?

By the time we get to the party, my kids have gotten fifteen gifts from well-meaning friends and relatives. So if that sounds familiar, tell your guests to bring an unwrapped book to donate in lieu of presents. The earth and someone who reads it will love you, even if your kids hate you.

Intriguing . . . I Can Handle a Little More

Start with an online e-card service like paperless post.com instead of paper invites and thank-yous. Use cloth tablecloths and napkins, and nonplastic decorations.

And there's no reason to white-flour- and sugar-up your guests just because someone has survived another year. I mean, you don't have to make a vegan cake with organic strawberries, agave syrup, and hemp flour just because I did, but at least avoid artificial colors and dyes in your frosting. And I know they don't sound fun, but vegetables, hummus, herbal iced tea, whole-grain pasta salads, and oven-baked chips all go over well at a party, especially my parties because people are just grateful there is something besides carrots to eat. Instead of soda, try a mix of fruit juice and seltzer. For candles, stay away from mainstream retail ones that are often made from petroleum by-products such as paraffin, which gives off the same fumes as diesel engines. Natural candles made of soy, beeswax, or palm oil are easy to find.

Activities and décor can also be low-key and eco-aware—paint T-shirts, turn recycled "junk" into sculptures or toys, or hold the party in a park and do a nature-based scavenger hunt. For older kids, do something meaningful, like a tree-planting, or something fun, like camping out in the backyard. Finally, no parent wants any more plastic crap cluttering their home, so choose party favors that are low-waste and useful, like puzzles, books, or seeds with little pots that kids can decorate.

I Need Some Facts to Bore My Friends With

If you're set on paper invites, try plantable wildflower paper. Make sure your cards include instructions on the back telling guests where to plant and how to care for their wildflower seedlings.

And if people are bringing presents, ask on your invitations that they be wrapped in recycled magazine pages or newspaper. Wastefully produced and possibly nonrecyclable, the wrapping just gets ripped to shreds in five seconds anyway.

Waste-Free/Junk-Free Lunches

Cut to the Chase, Hippie: What's the Least I Need to Know?

Even though bottled water and rice milk come in individual sizes, they produce unnecessary waste. Opt for reusable beverage containers like Klean Kanteen—they come in a variety of styles, and some are built to keep drinks cool for hours.

Intriguing . . . I Can Handle a Little More

It might be hard for your kid to resist the lure of the nuggets, hot dogs, pizzas, tater tots, and other fast-foodish woes of the cafeteria lunch line, so I like to step it up by making school lunches healthy and appealing. Some moms are better at "momming" than me and actually plan lunches with their kids for the week—something to aspire to that can make it easier to get your kids to eat healthier, although I imagine I still couldn't get them to eat that macrobiotic pho. Keep in mind that many kids want a lunch that looks like everyone else's. Whole-wheat pita sandwiches, wraps, and burritos are standbys.

It's easy to keep kids hydrated without sweet drinks (which do a poor job of hydrating anyway) by adding a slice of cucumber, some mint, or a lemon twist to the water bottle. And when it comes to the lunch box, look for a lead- and BPA-free or stainless steel variety that won't leach chemicals into the food. Instead of plastic bags, plastic wrap, or foil, use a reusable sandwich container and velcro resealable bags. Finally, throw in a cloth napkin and stainless steel cutlery.

I Need Some Facts to Bore My Friends With

In California, the Berkeley United School District's lunch program is perhaps the most innovative in the country—serving 5,500 healthy breakfasts and 3,000 lunches a day without a single preprepared food item on the menu. The district's food service employees prepare by hand and serve 4,000 pounds of fresh fruits, 700 pounds of beans, and 1,300 pounds of pasta every week. A great example for all school districts. Until your district makes the change, be the squeaky wheel. I did it—yes, I'm *that* parent.

School Buildings: Learning Where You Can Breathe

Cut to the Chase, Hippie: What's the Least I Need to Know?

Parents have rights—and responsibilities. Work with school authorities and join the parents group to clean up air quality and get a handle on chemicals being used in your kids' school. Make friends with the head custodian to get all the inside info.

Intriguing . . . I Can Handle a Little More

Lots of school buildings are old and in disrepair, teeming with big-time toxins like lead in the pipes or paint, and asbestos in the walls or ceiling. Then there's the possibility of molds and dust—major allergens. Say your school tries to spruce itself up with a new coat of paint or floor wax or varnish; well, okay, but what if it contains high volatile organic compounds (VOCs) that include cancer-causing toxic contaminants. Vermin an issue? What kinds of pesticides do they use, and how often? If you're lucky enough to have grassy or landscaped areas as part of the school complex, herbicides may be an issue. Just remember pesticides and herbicides have been linked to upper respiratory problems, cancer, liver and kidney damage, reproductive damage . . . Should I go on or are you ready to talk to your school? In case you need more motivation, those heavy-duty industrial cleaning solvents overflow with hard-core chemical irritants. The various chemicals in these cleaners pack quite a punch as they have also been linked to liver and kidney damage, reproductive damage, cancer, and blood disorders, among other things. Sound bad enough? Here's the part that slays me: Many cash-strapped schools attempt to cut down on heating and cooling costs by hermetically sealing buildings, trapping toxins in classrooms without any fresh air. It all adds up to a growing number of students and personnel complaining of headaches, respiratory issues, fatigue, dizziness, nausea, and trouble concentrating.

I Need Some Facts to Bore My Friends With

In 1995, the U.S. General Accounting Office estimated that 13 million kids attended schools with health-damaging environmental conditions (bad ventilation, lead exposure, pesticide issues, etc.). The result: Pediatric hospitalizations significantly increase in the days after summer, winter, or spring vacation ends, when kids return to schools suffering from Sick Building Syndrome.

clean rites
of passage

Tell me, what is it you plan to do with your one wild and precious life?

—Mary Oliver, bestselling poet

Cradle to Grave

I dread big life changes as much as the next person. I've been through many, some joyous, some less so. Good or bad, transitions are hard. I've made my way off to college, then into my first apartment, into relationships, and sometimes sadly out of them. I've moved many times (too many), and I've created a family. I wonder, how much extra stuff have I accumulated with each transition? What is it about change that makes us buy? Maybe it just gives us an excuse to spend, or promises a chance to redefine ourselves starting with the things that surround us. (Buying different junk certainly sounds easier than actually changing on some core level.)

Next time you face a big change, use it as an opportunity to define yourself as green—or at least light green—even if it means not buying quite so much. After all, everyone knows we are defined by the inside out. And maybe by one amazing pair of boots.

Off to College: Sometimes We Have to School the Administration

Cut to the Chase, Hippie: What's the Least I Need to Know?

Bringing your own place setting to the cafeteria can prevent your weight in plastic waste each year. So shell out the couple bucks for a thrift-store mug, bamboo plate, and dollar-store flatware. If you're going to pack on the "freshman fifteen," might as well do it sustainably.

Intriguing . . . I Can Handle a Little More

College campuses produce a mind-boggling amount of waste. Everything from Sheetrock to water bottles to broken MacBook Pros—in the race for championing state-of-the-art facilities, campuses are sites of rapid demolition and construction and environmentally costly green space maintenance. Dominated by a philosophy of convenience, students and faculty are encouraged to eat and run, chucking mountains of polystyrene plates and plastic containers on their way out the door. And the carbon footprint is deep. Classroom buildings keep their track lights buzzing all hours of the night, and climate-controlled dormitories auto-blast boilers all winter long. I know this firsthand, since one kept me up all night long every night my senior year, rattling and whirring. And yes, I made the brilliant choice to live in a basement dorm directly next door to the boiler room. Join your school action group and insist on change.

I Need Some Facts to Bore My Friends With

In 2000, the average college student produced nearly 640 pounds of waste each year, including 320 pounds of bond paper for printing all those essays on Foucault and economic theory. At the time, the leading campus food supplier was Sodexo, most noted for their accounts with nearly every American penitentiary, but pressure from students has put a dent in Sodexo's near-monopoly and other environmentally unfriendly deficiencies. For the last several years, American colleges have been pitted against one another in a race for a new kind of prestige: greenest campus. Between 2006 and 2007, five hundred campuses created student-and-faculty-facilitated sustainability groups, and the United States Green Building Council has begun subsidizing universities that commit to building carbon-controlled classrooms, using solar panels, LED lighting, green windows, and gray-water systems (where laundry, bath-, and sink water runoff is recycled to water gardens).

From Pratt Institute in Brooklyn to University of Washington in Seattle, undergraduates have been at the forefront of campus composting committees, vegetable oil collection for biofuel-run campus vehicles, free bicycle clinics, and petitions for LEED-certified construction. On campuses where these measures are active, nearly seven hundred pounds of food and three hundred pounds of coffee grounds are converted into fertilizer for campus property and nearby community gardens. Part of this movement has included efforts to link community sustained agriculture farmers with college kitchens.

Setting Up Your First Green Apartment

Cut to the Chase, Hippie: What's the Least I Need to Know?

When you live in a building where you don't control what kind of heat you get, you can still opt to crank it down. Turning your thermostat down by just two degrees in the wintertime will prevent over two tons of carbon dioxide from saturating the atmosphere. Too cold? Get another blanket or try sleeping with a hot water bottle. Just don't make yourself sick over it.

Intriguing . . . I Can Handle a Little More

When you're looking for your own place, choose a neighborhood where you won't have to drive to stock up on all of your essentials—eye the walking distance from your nearest grocery or convenience store. You may also be able to negotiate a deal on your rent by offering to do some kind of work trade—like being responsible for bringing recycling and garbage to the curb every week.

One of the most difficult and energy-sucking aspects of apartment life is the lack of decent lighting. Opt out of midcentury buildings, which tend to have shallow ceilings, smaller windows, and dingy or insufficient ceiling fixtures. Look for units above ground level. Find a couple of second-hand floor lamps, and hang lightweight drapes instead of the toxic vinyl venetian blinds. At night, eat by candlelight. For a healthier glow, choose vegetable-based candles, which unlike their wax counterparts are petroleum-free and won't leave those spooky smoky trails on your walls. They last for up to twice as long, which is good news for those of us who don't have cash to burn.

Furnishing your first apartment is an incredibly personal task, like narrowing down an outfit you're happy to wear every day for the next calendar year. That said, some of the best finds might just be in thrift stores, on the street corner, or in Craigslist's "Free" postings.

When you're moved in, stock up with a few hand-mixed, vinegar-based cleaners, and one oil-based for wood floors. Keep your electronics plugged into a power strip so you can turn everything off at night with one effective click.

I Need Some Facts to Bore My Friends With

Think you have to move to a biodome in Vermont or a house made of trash in New Mexico to be ultragreen? Think again. Cities are actually greener. The average Manhattanite apartment dweller consumes gas at rates akin to those prior to World War I. And Los Angeles tenants, who certainly drive their fair share, use half as much electricity than any other metro area on the West Coast.

Clean Dating: Why Diamonds and Roses Aren't Always So Romantic

Cut to the Chase, Hippie: What's the Least I Need to Know?

Smell the roses: If a rose has no scent, it's likely been genetically modified or chemically treated.

Intriguing . . . I Can Handle a Little More

I've been there. You're sweating, your heart is pounding fast. Gotta bring a gift. The standard thing to do is run into the grocery store and grab some flowers. But ironically, the most traditional romantic gifts—roses and diamonds—are often harvested and mined in ways that harm the planet as well as workers' health. Know where your gifts come from. You're safest buying local flowers and vintage gems. Or do something different: One of the sweetest gifts anyone ever brought me was a homemade vegan pizza.

Gifts aren't the only thing to think about when we're dating, of course. Finding partners who share our environmental consciousness can be as important as smarts and a sense of humor. Several online dating services match "green" singles—whether you're looking for a fellow animal-rights activist, a vegetarian, or simply someone who identifies as green, you can be sure none of these folks will show up to your first date driving a Hummer.

I Need Some Facts to Bore My Friends With

They say roses symbolize love, but by the time those vibrant flowers reach your trembling hands, most will have been treated with a battery of potentially lethal chemicals.

Watch out for all imported blooms. Colombian flower workers have suffered mass poisonings, and children in flower-growing regions of Ecuador have shown developmental delays because of the toxins their mothers were exposed to during pregnancy. That's not my idea of love.

When it comes to jewel mining, the process isn't inherently destructive, but large-scale operations often involve a lot of bloodshed and the dumping of massive amounts of harmful waste into water systems around the world.

Still, industries change. In response to consumer outcries, the diamond industry has done a lot to clean up its act in recent years. Keep the pressure on gem dealers by insisting on proof of a product's history. If you choose a new diamond, make sure it's certified "Conflict Free." Better yet, choose a locally crafted piece of jewelry or a vintage gem—and personalize it with a new inscription.

who knew? oceans

The oceans are in trouble, which means we are, too. This is thanks in part to widespread development, unsustainable fishing, global warming, marine hunting, and deep-sea mining. But the big story here is marine pollution. On the grand scale, oil spills are dangerous, but everyday pollution is most common. Need proof? The most stupefying example is the Great Pacific Garbage Patch, which takes up an estimated 8 percent of the world's biggest ocean. This garbage ultimately breaks down into toxic particles or gets ingested by birds and fish that mistake it for food.

Weddings and Commitment Ceremonies (No, She Won't Wear It Again)

Cut to the Chase, Hippie: What's the Least I Need to Know?

Some 2.5 million weddings and receptions are held each year—and all those guests use about a gazillion barrels of oil to get there. The eco-trick here isn't complicated: If you're planning to tie the knot, choose a location that the fewest number of people need to travel to.

Intriguing . . . I Can Handle a Little More

Between fuel, food, and partying, twenty tons of carbon (a cause of global warming) and $20,000 are expended for the average American wedding. It hasn't always been this way. The wedding industry began amping up expectations for this "perfect day" in the roaring, pre-Depression 1920s. So while getting hitched may be worthy of a big deal, there's no need to be wasteful in the name of love.

Hold your ceremony at home, at a state park, or look for cheap venues like community centers, or American Legion halls—they almost always have a yard, a barbecue pit, a rec room, and a bar—all for about $300.

Make a list of the things that are most important to you—say, drinking, eating, and dancing—and allocate your energy and money to those personal basics. Let the bridesmaids wear their own clothes, offering them color or other guidelines. Think this is a bad idea? Don't kid yourself—whatever they have is better than any bridesmaid dress you'll pick. Think your taste is the exception? It's not.

Glean your community for musicians, photographers, designers, cooks, cake decorators, and tech-savvy folks and put them to work. Most people won't just be willing to help, they'll be honored. Keep the menu simple. Opt for a family-style meal since buffets tend to result in pounds of unwanted food. See to it that leftovers are donated rather than disposed of, or delegate someone to cart compost over to a needy community farm. Buy your alcohol by the gallon, and serve from kegs or barrels into nondisposable cups.

For invitations, go for paperless e-vites. If you're too sentimental for online invitations, look for recycled stationery and soy-ink printers. Grab your flowers from a farmers market, and stick to the seasonal stems. And don't give out schwag you wouldn't keep yourself. Seeds make a great favor—and even if they don't, your guest who drank all those gallons of beer from the keg won't know the difference.

I Need Some Facts to Bore My Friends With

The American wedding industry is now a $160 billion business, and not much of it's green. At one factory in Xiamen, China, skilled migrant seamstresses who live eight to a room turn out 100,000 wedding dresses every year. Those dresses sell for an average of $1,025 each. Guess how much those skilled seamstresses make. Yep: six dollars a day.

Divorce and Child Custody: Staying Green Through the Transitions

Cut to the Chase, Hippie: What's the Least I Need to Know?

Instead of picking up a storage unit where you're keeping all the crap neither of you can decide what to do with, donate shared belongings.

Intriguing . . . I Can Handle a Little More

Whether your divorce is happy, contentious, or an overwhelming mixed bag, separation usually produces two households where there was one—each with nearly identical carbon footprints as the original primary household. We're talking double the heat, water, and gas consumption, and perhaps double the cars, TVs, and frying pans, for the same number of people. Consider taking the opportunity to downsize and simplify your life as you make your way through the transition.

When shared child custody is part of the picture, consider remaining within the same zip code (or thereabouts) to save gas and mental health—being within walking or bussing distance from each other keeps you from long, fuel-heavy commutes shuttling children and forgotten backpacks, and can also prevent the parents' worlds from seeming totally detached from each other. Keeping in close proximity to the same schools, neighborhood friends, and other familiar locales can help balance other disorienting aspects of a changing family.

I Need Some Facts to Bore my Friends With

If lawyers need to be called in, save money and heartache by employing a lawyer who specializes in holistic divorce—with a background in mediation, psychology, or some other form of social work outside of legal experience. Yep, there are actually divorce lawyers out there who know how to prevent you from destroying each other.

who knew? wildfires

Scientists believe that as the earth gets hotter and drier, wildfire seasons will get longer and the fires bigger. Modern fire-management systems utilize controlled burning, but this can go horribly wrong. Many environmentalists disapprove of prescribed burns, especially if wildlife conservation is not factored in. Don't think of wildfires as someone else's problem—even if you're hundreds of miles away, the plume can linger, spewing ash and pollutants and causing illness and respiratory troubles. Try writing to your elected officials against controlled burn and, of course, do your part to assuage climate change.

Green as You Gray: Eco-Conscious Retirement Communities

Cut to the Chase, Hippie: What's the Least I Need to Know?

Like any other industry, the retirement community market is full of false advertising. Unless you're looking at medical-attended assisted living, most facilities are unregulated and all of them are businesses, which are free to set whatever costs they see fit. Look for options that stress simplicity over sustainable spectacle.

Intriguing . . . I Can Handle a Little More

Many facilities will make claims about their efforts to protect natural resources and boast of highfalutin green components without any results or certification to show for it. While it's important to eye the details, low-flow faucets and microclimate landscaping don't mean much if half the eco-friendly efforts go into the golf course's water-recycling technology. There's nothing wrong with institutional recreation, but if you're in the market for some form of facility living minus the energy suck, look for low-impact living with the least amount of eco-jargon. Let's face it: Access to public transportation goes a lot further than a state-of-the-art HVAC system.

I Need Some Facts to Bore My Friends With

If you're interested in institutional retirement, look for sites reminiscent of college campuses, complete with communal buildings. You'll be walking distances to food, gyms, and cultural centers. It's also worth checking to see that your campus is LEED-certified—it's a regulated way to know that the building you're living in is holding itself accountable for its carbon footprint. And while compact fluorescent lightbulbs, customizable utilities, and Energy Star appliances make their contribution, the gadgets aren't the end-all-be-all. The sanctity of a green retirement community lies in its overall form.

If nonassisted living is more your cup of tea, consider senior cohousing or cooperatives, which often aren't age segregated but allow you to retire in a community where no one is left in isolation. I mean, who doesn't need a green shoulder to lean on?

Let's Finish What We Started: Green Burial and Cremation

Cut to the Chase, Hippie: What's the Least I Need to Know?

The most eco-conscious way to go: Green cemeteries and memorial nature reserves have flexible policies for low-impact burials.

Intriguing . . . I Can Handle a Little More

If what's important to you and your loved ones is a green ending, an eco-burial is always going to be less taxing on the environment than a cremation. There's the pollution factor: Methane, carbon dioxide, and even mercury from dental fillings all contribute to air quality. If cremation is the way you're going, you can lessen the environmental impact by requesting that the body go into the furnace in an unlined, nonsynthetic container. Green crematoriums also use double burners and can discuss measures they're taking to limit pollutants.

More expensive but greener, a natural or green cemetery fosters a friendly habitat for native birds and mammals, and the groundskeepers replenish and germinate native flora and grasses. Each venue does this in a variety of different ways, and it's important to ask questions. There's the option of a Biopod, which is a specially manufactured coffin that decomposes within 180 days, and allows the body to enter the earth quickly—but most of these places offer a simple box (made of natural fibers), or even wrapping in a special blanket and putting the body straight into the earth, no embalming fluids or anything—and because those services are removed, almost any household can afford it.

I Need Some Facts to Bore My Friends With

Human bodies are made to biodegrade. And expensive embalming fluids, sealed caskets, and memorial vaults aren't mandated by law. The thought of decomposing freak you out? It's going to happen anyway, it just takes longer with all that embalming gunk. Get over it—ashes to ashes, dust to dust, baby.

acknowledgments

A big thank-you to everyone who supported me emotionally and creatively through the process of putting this book together (not a small undertaking): Lori Sunkin, Collier Schorr, Nina Garduno, Jenny "Esther" Brill, Jonah Wilson, Leonardo DiCaprio, Steven Levy, Dr. Steven Glass, Shannon Del, Lisa Kudrow, Manuel Nieto Jr., and Ana Maza.

Thanks to my work family for being in my corner for this book and everything else: CBS, Les Moonves, Nina Tassler, CBS Studios, Glenn Geller, Angelica McDaniel, and all of my friends at *The Talk* who make my life richer and let me prattle on about possible titles, segments, etc., including Sharon Osbourne, Julie Chen, Aisha Tyler, Sheryl Underwood, John Redmann, Kristin Matthews, Heather Gray, Billy Bowers, Ann Marie Williams-Gray, Masumi Ideta, Gaye Ann Bruno, Kelli Raftery, Erin O'Brien, Jennifer Solari, Ann-Marie Oliver, Cheryl Eckert, and the entire staff and crew.

I am truly grateful to the people who are smarter than I am and helped me where I needed it: Tammy Schwolsky, Dr. Jay Gordon, Nadine Barner, Annie Moizan, Bibi Deitz, and Adrian Shirk.

A special thank-you to those of you without whom this book would not exist: Ali Adler for setting the bar creatively and always providing a counterpoint to my beliefs (therefore pushing me to be strident enough to actually write a book); my astonishingly smart, funny, and tireless researcher, Ariel Gore; Amy Lederman (this is all your fault); Pamela Cannon, Ratna Kamath, Betsy Wilson, and everyone at Ballantine Books for much patience, guidance, and belief in me; the Gersh Agency, Todd Christopher, and Joe Veltre for making it happen; and of course, my amazing mom, Barbara Gilbert-Cowan, without whom I wouldn't exist—so if you don't like the book, blame her.

And thank you to LP for loving me through it all and inspiring me with your brilliant talent and unique voice.

index

a

acidic diets, 16
acidosis, 10
activism, environmental, xiii–xiv, 114, 115
acupuncture, 101, 102
A Divine H$_2$O, 6
adoption, 152
advertising, 55
Affordable Health Care Act, 158
air conditioning, 39
air pollution
  in cities, xv, 110
  home air purifiers, 101
  indoor, 55
  toxins in cleaning products, 59
  transportation-related, 128, 136, 137
air travel, 144
alcohol-based sanitizers, 54

alcoholic beverages, 23
alkaline copper quat, 47
alkaline diets, 16
alkaline filters, 6
allergies
  essential information, 101
  formaldehyde exposure and, 102
  soy, 17, 19
  treating, 102
Alliance for Justice, 115
aloe vera, 95, 103, 156
aluminum, 69
Alzheimer's disease, 22
American Apparel, 97, 130
American Community Garden Association, 113
American Society of Home Inspectors, 33
American Solar Energy Society, 126

American Wind Energy Association, 126
Amish, 116
ammonia, 65
animal cruelty
  in apparel industry, 97
  in circuses, 167
  in meat industry, 10, 72
  pest control without, 56
antimony, 72
antioxidants, 11, 15, 24
apartment buildings, 177
apples, 7
aquaculture, 12, 13
arachidonic acid, 159
Archer Daniels Midland, 127
arnica, 103, 156
arsenic
  in cigarettes, 105
  in drinking water, 6
  in treated wood, 166
arthritis, 22

asbestos, 42
aspirin, 103
assisted living, 181
attachment theory, 162
Aubrey Organics, 95
Audubon, John James, 150
avobenzene, 95
ayurvedic medicine, 116

b

Baby Buddy, 94
bamboo products, 26, 35
Ban the Box campaign,
    xiii–xiv
barium, 72
bathrooms, 57
batteries
    car, 139
    computer, 67
bedding materials, 53,
    153
BeerTown.org, 23
bees, 80
beet sugar, 25
Begley, Ed, Jr., 38
beriberi, 104
berries, 7
Berry, Wendell, 116
bicycling, 38, 100, 117, 118,
    128, 142
Bikram yoga, 98
Billion Tree Campaign, 110
biodynamic agriculture, 23
biofilms, 54
biofuels, 25, 138
Biopod, 182
Birds of America, The
    (Audubon), 150
birthday celebrations, 168

Bisphenol-A (BPA), 6, 71,
    94, 151, 153, 159
bleach, 65
blueberries, 7
books, 62
Borax, 60
Botany of Desire, The
    (Pollan), 4
BPA. *see* Bisphenol-A
Bradley method, 155
breastfeeding, 156, 157,
    158
brewpub locator, 23
brown rice syrup, 25
Buddhism, 116
Buffalo Exchange, 111
Buffet, Warren, 30
Building Performance
    Institute, 33
burial practices, 182
butterflies, 19
butylcellosolve, 59

c

calendula, 103
California Baby, 95
Campaign for Safe
    Cosmetics, 92
candles, 168, 177
cane sugar, 25
carbon dioxide, 128,
    136
carbon footprint, 136
carbon monoxide, 33,
    136
carbon offsets, 144
carpets and rugs, 35, 41
    cleaning, 61
carpooling, 118

cars, 128, 135
    alternative
        transportation,
        140–43
    biofuel-powered, 138
    environmental harms of,
        128, 136
    flexible fuel vehicles,
        138
    strategies for reducing
        emissions from, 137
Carter administration, 43
cataracts, 22
cell phones, 68
cemeteries, 182
Center for Science in
    Public Interest, 115
Centre for Sustainable
    Fashion, 97
Chicago, Illinois, 142
chicken pox, 160
chicken(s), 9, 84
childbirth, 154, 155
China Study, 14
chlorine, 46
chlorine bleach, 65
chloroform, 6
chocolate, 24
Christianity, 116
Christmas, 72
chromium copper arsenate,
    166
chrysotile, 42
Churchill, Winston, 122
cigarettes, 105
cinnamon, 22
circuses, 167
citronella, 103
Clean Air Act, 59
Clean Edge, 131
Clean Water Act, 96

climate change and global warming
  community design considerations, 117
  current understanding of, 8
  effects in polar regions, 164
  food waste and, 9
  threat of, 8, 128
  wildfire risk, 180
  see also greenhouse gases
closed loop manufacturing, 70
clothing
  baby, 153
  buying used, 111, 130
  environmental considerations, 97
  sweatshop labor in making of, 130
  wedding dresses, 179
cocamidopropyl betaine, 94
Coconut Bliss, 14
cod, 12
coffee, 21
coffins, 182
colds and flus, 54
college life, 176
Colony Collapse Disorder, 80
comfrey, 103
community action, 109
  on college campuses, 176
  consumer protection, 115
  in faith communities, 116
  gardens, 113
  giveaway and free shops, 112
  opportunities for, 118

for policy reform, 114
  tree planting campaigns, 110
Community Environmental Legal Defense Fund, 114
community sponsored agriculture, 8, 164
commuting, 128
composting, 78
  biodegradable plastics, 71
computers, 67
concrete flooring, 35
consumer advocacy, 115
Consumers Union, 115
Consumer Watchdog, 115
Continental Airlines, 144
Cool Roof Rating Council, 36
cool roofs, 36
cork flooring, 35
corn, 10. *see also* high–fructose corn syrup
cosmetics and beauty products, xiv, 91
  environmental and health concerns, 92
  essential information, 92
  hair care, 93
cotton, 97, 153, 161
cremation, 182
Crossroads Trading Company, 111
cutting boards, 26
cyanazine, 153
cyanide, 105

d

dairy products, 14
Daiya, 14

dams, 127
Database of State Incentives for Renewable Energy, 126
dating, 178
Davis, Ms., xiii
decks and terraces, 47
Deen, Paula, 84
deforestation, 10, 104, 143
DeGeneres, Ellen, 50
dental health, 94
Department of Energy
  Building American Program, 34
  Energy Efficiency and Renewable Energy website, 44
detoxification diet, 20
Devine Color Paint, 55
Devita, 92, 95
diamonds, 178
diapers, 161
diatomaceous earth, 56
dibutyl phthalate, 96
dicofol, 153
dioxane, 64
dioxins, 71, 161
direct-trade products, 24
dishwashing, 60
disinfectants, 6, 12
divorce, 180
docosahexaenoic acid, 159
doula, 155, 156
downcycling, 70
drugs, prescription. *see* pharmaceutical drugs
dry cleaning, 97
DuChateau flooring, 35

e

E. coli, 167
Earthborn Paints, 55
Earth Mama Angel Baby, 156
Earthpaint, 47
earthships, 38
echinacea, 103
Economic Development Directory, 126
economics
    advantages of buying locally grown products, 8
    advantages of vegetarian and vegan diets, 13
    benefits of living simply, 124
    dehumidifier energy efficiency, 40
    electric cars, 139
    fair labor practices, 130
    fair trade products, 21, 24, 25
    of GMO foods, 19
    green home design and construction, 34
    harms of big box stores, 18
    home heating and cooling, 39
    investing, 131
    solar energy, 32
    spending on fast food, 11
    spending on weddings and receptions, 179
    washing laundry, 64
eco-roof, 36
Ecos Paints, 55
eco-trade products, 24

eCycling program, 93
Edinburgh Postnatal Depression Scale, 156
electric and hybrid vehicles, 135, 137, 139, 141
electric cigarettes, 105
electricity
    geothermal, 66
    from human motion, 100
    LED lights, 41, 72
    nuclear-generated, 131
    off-grid living, 38
    solar-generated, 43
    wind-generated, 44
electric vehicles. see electric and hybrid vehicles
electronic devices
    computers, 67
    energy consumption, 62, 66
    recycling, 93
endangered species, 32
    fish, 12
    global warming effects, 164
endocrine disruptors, 6, 10, 52, 95
Energy Star label, 40, 64, 67
energy use
    alternative fuels market, 131
    in apartment living, 177
    audit, 33
    benefits of green home design, 34
    benefits of living simply, 124
    benefits of weatherizing, 37

    in electronic device manufacturing and operations, 62, 67
    fracking for natural gas, 43
    grants to reduce, 126
    for heating household water, 64
    for home heating and cooling, 37
    house design and operations to reduce, 32, 34
    humidifiers and dehumidifiers, 40
    lighting, 41
    off-grid living, 38
    in paper making, 129
    televisions, 66
    for transportation, 128
    wind power, 44
    see also electricity; solar energy
environmental awareness
    challenges in everyday living, xiii–xvi
    children's, 163
    in household, 51
Environmental Protection Agency, 46, 59
    asbestos website, 42
    eCycling program, 93
    Electronic Product Environmental Assessment Tool, 67
    lead abatement regulations, 45
    list of cell phone recyclers, 68
Environmental Science and Technology, 100

Environmental Working Group, 6, 92, 94, 151
epidural injections, 155
e-readers, 62
erosion, 82
essential fatty acids, 159
essential oils
   for fragrance in body-care products, 54, 95, 103
   pest control with, 56
   vaporizer use, 101
estrogen, 17
ethanol, 25, 138
exercise and fitness, 98
   electricity generation through, 100
   hiking, 100
   Tai Chi and Qigong, 99
   yoga, 98

f

fabric cleaner, 61, 65
factory farming, 10
fair-trade products, 21, 24, 25
Falkland Islands, 164
farmers' markets, 8
farming
   aquaculture, 12, 13
   biodynamic, 23
   cattle, 157
   coffee, 21
   cotton, 97, 161
   factory, 10
   genetically modified organisms, 19
   monoculture, 10

sugar beets, 25
   urban, 86
fast-food restaurants, 11
fat in diet, 11, 12
fennel, 22, 156
fertilizer, 80, 82
   compost, 78
filters
   air conditioning, 39
   air purification systems, 101
   tap water, 6
Firemaster 500, 52
fireworks, 72
fish, 12, 13, 151
five flower formula, 103
flame retardants, xiv, 52, 67
flea control, 63
Fletcher, William, 104
flies, 56
floors and flooring material
   carpet, 35
   cleaning products, 59
   green sources, 35
   reclaimed, 34
flowers, 178
fluorescent bulbs, 41
fluoride, 94
food
   college meals, 176
   dining out, 9
   juicing, 20
   kitchen tools, 26
   locally grown, 8
   meal preparation as family event, 165
   nutrient needs of children, 164
   organic, 7
   preparation of nonorganic produce, 7

raising chickens for, 84
   raw, 15
   relative environmental impacts, 9
   for wedding receptions, 179
   see also gardening; vegan diet; vegetarian diet; specific food or food type
Food and Drug Administration, 92
Food Policy Journal, 8
Forbes Magazine, 125
Forest Stewardship Council, 35, 47, 166
formaldehyde, 52, 55, 96, 97, 105
Foundation Center, 126
fracking, 43
Freecycle network, 112
free stores, 112
frogs, 10
Fukishima tsunami, 131
furnaces, home heating, 39
furniture
   bedding materials, 53
   for children, 153
   cleaning products, 61
   outdoor, 47
   toxic threats in, 52

g

gallstones, 13
gardening
   benefits of, 77
   community, 113
   compost fertilizer for, 78
   in containers, 80
   fertilizer use, 80, 82

gardening (*cont.*)
  organic practices, 79
  pest control, 82
  rooftop, 36
  soil testing, 79, 80
  urban, 86
  water use, 83
genetically modified
    organisms, 127
  health risks, 19
  soy products, 17
geothermal energy, 66
geotourism, 145
ginger, 22, 103
giveaway shops, 112
glycol ethers, 59
glyphosate, 82, 127
GMOs. *see* genetically
    modified organisms
goats, 85
goldenseal, 103
Gore, Al, 144
grants, green, 126
Green Bible, 116
Greencradle, 153
Green Dream Jobs,
    125
greenhouse gases, 136,
    157
  carbon footprint
    calculation, 137
  dairy industry as source
    of, 157
  meat industry as source
    of, 10
  transportation-related,
    13, 23, 128
green industries and jobs,
    125
Green Microgym, 100
Green Restaurant
    Association, 9

green roofs, 36
Greensburg, Kansas,
    117
Green Sprouts, 94
Green toys, 153
greenwashing, 55
grindelia, 103
*Guinness Book of World
    Records,* 63

h

habitat preservation, 32
hair care, 93
Halal foods, 10
halibut, 12
hand cleaners, 54
hatha yoga, 98
heart disease, 11, 13
heating and cooling
  apartment living, 177
  benefits of
    weatherizing, 37
  essential information, 39
  geothermal, 66
heat pumps, 39
herbal remedies, 103
herbal teas, 22, 156
herbicides, 82, 170
hexavalent chromium, 67
high-fructose corn syrup,
    11, 25
hiking, 100, 143
Hindu religion, 116
holidays and
    celebrations, 72
  children's parties,
    168
homeopathy, 101
hormones
  in food for farmed

fish, 12
  see also endocrine
    disruptors
Hourglass cosmetics, 92
house design and
    construction
  bathroom, 57
  flooring materials,
    34, 35
  floor space, 32
  grants for, 126
  green, 34
  inspection, 33
  for off-grid living, 38
  outdoor decks, 47
  renovating older houses,
    31, 41, 42, 45
  roofing, 36
  site considerations,
    32, 34
  for solar power, 43
  use of salvaged and
    recycled materials in,
    34, 38
  water catchment
    systems, 36, 43
  weatherizing and
    insulating, 37, 39
  windows, 37
  see also house operations
    and maintenance
house operations and
    maintenance
  for allergy management,
    101
  cleaning products, 59
  dishwashing, 60
  heating and cooling,
    39, 66
  humidity control, 40
  pest control, 56
  swimming pools, 46

see also house design and
    construction
Human Rights
    Watch, 10
humidifiers and
    dehumidifiers, 40
hydrocarbons, 136, 151
hydroelectricity, 127
hydrogenated fat, 11
hypertension, 15

i

ice cream, 14
Iceland, 66
incandescent bulbs, 41
insecticides, 56, 82
insulation. *see* weatherizing
    and insulating
investing, 131
ionizing filters, 6
iron, dietary, 13, 164
Iroquois Confederacy, 148,
    150
Islam, 116
isoflavones, 17
Iyengar yoga, 98

j

JASON, 93, 94, 95
jewels, 178
jewelweed, 103
John Masters
    Organics, 95
Judaism, 116
Juice Beauty, 92
juices, 20
Junior Ranger program,
    163

k

kapok, 53
Katrina, Hurricane, 52
Keen, Jim, 167
Kickstarter, 126
kidney stones, 13
kitchen tools, 26
kundalini yoga, 98
Kyoto Protocol, 117

l

labeling and certification
    cool roofs, 36
    cosmetics and beauty
        products, 92
    dehumidifier energy
        efficiency, 40
    greenwashing
        practices, 55
    non-GMO foods, 19
    organic foods, 7, 19
    rain forest
        protection, 24
    recyclable plastics, 71
    vitamins and
        supplements, 104
    window efficiency, 37
    wood products, 35
La Leche League
    International, 157
Lamaze method, 155
Lamott, Anne, 90
landscaping, 86
laundry, 64, 65, 97
lavender, 63, 103
lawyers, 180
lead
    in beauty products, 92
    in drinking water, 6
    in paint, 45
    in pipes, 45
Leadership in Energy and
    Environmental Design
    (LEED) rating, 126
leather goods, 61, 97
LeBoyer Gentle Birth,
    155
LED lights, 41, 72
LEED rating. *see* Leadership
    in Energy and
    Environmental Design
    rating
leftovers, 9
libraries, 62
lighting and light fixtures,
    41, 72, 177
linens, 65
linoleum, 35
lipstick, 92
local building
    materials, 34
localharvest.org, 8
locally grown food, 8
locally owned
    businesses, 18
London, Stacy, 93
Lorax, The (Seuss), xi
Los Angeles City
    College, 38
lye, 93

m

mackerel, 12, 151
macrobiotic diet, 5, 16
makeup, 92
manicures, 96
maple syrup, 25
Marine Stewardship
    Council, 12

mattresses, 53
Mayo Clinic, 13
Mayors Climate Protection
    Agreement, 117
meadowsweet, 103
meat
    dining out, 9
    environmental harms of,
        10, 13, 136
    essential information, 10
    good sources of, 10
    industry practices, 10
    soy-based
        imitations, 17
medicinal plants, 103
meditation, 98
mercury, 12, 67, 160
methane gas, 9, 13, 136
methylcycloprene, 7
mexoryl, 95
mice, 56
midwifery, 154
migraines, 22
milk, 14, 157
    bottle-feeding babies,
        159
    goat, 85
Milne, A. A., 76
minerals in water, 6
mint, 56
miso, 17
moisturizing lotions, 95
monoculture farming, 10
Monsanto, 19, 127
Mormon church, 116
morning sickness, 22
motorcycles and scooters,
    141
Muir, John, 108
mulching, 82
Muslim food
    preparation, 10

n

Nader, Ralph, 115
nail care products, 96
naled, 153
National Association of
    Home Builders, 32
National Fenestration
    Rating Council, 37
National Geographic, 145
National Lead Hotline, 45
National Park Service, 163
National Research
    Council, 14
National Resources
    Defense Council, 128
Natural American Spirit,
    105
Nature's Gate, 94
Nelsons Hemorrhoid
    Cream, 156
neonicotinoids, 80
netbooks, 67
neti pot, 101
Neurotoxicology, 35
New Economics
    Foundation, 8
New York City, 142
nitrogen oxide, 136
nitrosamines, 95
nonylphenol
    ethoxylate, 64
norovirus, 54
nuclear energy, 131

o

Obama, Barack, 166
obesity, 11, 13
ocean pollution, 178
Oliver, Mary, 174

omega-3 fatty acids, 12
omnibus, 140
Only the Super-Rich Can
    Save Us (Nader), 115
Oral-B, 94
organic cigarettes, 105
organic cotton, 97
organic food
    benefits of, 7
    coffee, 21
    essential information, 7
    growing, 79
    labeling requirements,
        7, 19
    during pregnancy, 151
    transportation costs, 8
    wines, 23
Osbourne, Kelly, 93
oxybenzone, 95
oxytocin, 157

p

packaging, 18, 71
paint
    essential information, 55
    lead-based, 45
    outdoor deck, 47
palmetto bugs, 56
paper
    environmental
        considerations, 62, 129
    for invitations, 168, 179
    recycling, 69
parabens, 94, 95
parenting
    adoption, 152
    attachment theory, 162
    bottle-feeding babies,
        159
    breastfeeding, 157, 158

childbirth, 154, 155
child's exposure to
    animals, 167
divorce and child custody
    issues, 180
environmental
    responsibility and, 149
fostering child's
    connection to food
    production, 8
fostering child's
    environmental
    awareness, 163
hosting children's parties,
    168
meal preparation as
    family activity, 165
nursery furnishings, 153
pregnancy, 151
school lunch preparation,
    169
vaccination issues, 160
vegetarian families,
    164
particleboard, 52
patio furniture, 47
PBDEs, 52, 151
Peden, James, 63
penguins, 164
peppermint, 56
perchlorates, 72, 151
perchloroethylene, 61
pertussis, 160
pest control, 82
pesticides, 24, 56, 80, 82,
    97, 151
    advantages of organic
    foods, 7
    in food for farmed
    fish, 12
    in milk, 12
    use in schools, 170

pets
    acupuncture for, 102
    essential information, 63
    restaurant leftovers for, 9
petting zoos, 167
PFCAs, 11
pharmaceutical drugs
    allergy treatment, 101
    in childbirth, 155
    disposal, 58, 96
    in food for farmed
    fish, 12
    natural alternatives, 103
    water contamination
    with, 6, 58, 96
phenols, 59, 151
phosphates, 60
phthalates, 35, 71, 94, 95,
    151
Phyo, Ani, 15
phytic acid, 17
plastics
    children's exposure to,
    153, 159
    diapers, 161
    environmental concerns
    with, 71
    food handling and
    storage products made
    of, 26
    in playground
    equipment, 166
    recycling, 71
    shopping bags, 18
    water bottles, 6
playground equipment,
    166
Plug In America, 139
polar bears, 164
political action, 114
    consumer rights
    advocacy, 115

Pollan, Michael, 4, 10
polycarbonate plastic, 6
polychlorinated biphenyls,
    151
polycyclic aromatic
    hydrocarbons, 151
polypropylene, 71
polystyrene, 71
polyvinyl chloride (PVC),
    35, 37, 47, 71, 94,
    159
population growth, 152
Porchon-Lynch, Tao, 98
Portland, Oregon, 142
postpartum care, 156
potlatch, 112
potting soil, 80
PPL, 127
pregnancy, 151
    childbirth, 154, 155
    postpartum care, 156
Priestly, Joseph, 150
propargite, 153
protein, dietary, 14, 164
Public Citizen, 115
public transportation,
    140
PVCs. see polyvinylchloride

q

Qigong, 99
quercetin, 101
quinoa, 164

r

radon testing, 151
Rainforest Alliance, 24
rain forests, 10, 24, 104

rainwater catchment
  systems, 36, 43, 83
rapidly renewable
  materials, 35
raspberry leaf, 156
raw food diet, 15
Reagan administration, 43
Really, Really Free Market,
  112
recycling and salvaging
  benefits of, 69
  electronic appliances, 93
  house construction
    materials, 34, 38
  house furnishings, 41
  opportunities for, 69
  paper, 129
  plastic materials, 71
  preparation of materials
    for, 69
  roofing materials, 36
  telephones and cell
    phones, 68
  upcycling and
    downcycling, 70
religious organizations,
  116
rennet, 14
rennin, 14
Rescue Remedy, 103
respiratory problems, 22
restaurants and dining
  out, 9
  fast-food
    establishments, 11
retirement, 181
reverse osmosis filters, 6
Reynolds, Michael, 38
rheumatoid arthritis, 15
rice, 104
Ringling Bros. and Barnum
  & Bailey Circus, 167

roaches, 56
roof materials, 36, 83
rooftop gardens, 36
rubber roofing, 36
Rubio Monocoat, 35

S

Safe Drinking Water Act, 6
salmon, 12, 13
San Diego, California, 142
sanitizers, hand, 54
Sante Fe, New Mexico,
  110
schools
  building health, 170
  lunches, 169
seafood, 9, 12
Sears, William, 162
seaweed, 14, 78
senior housing, 181
septic systems, 57
Seuss, Dr., xi
Seventh-Day Adventists,
  116
Seventh Generation
  concept, 148, 150
Seventh Generation
  products, 161
shade-grown coffee, 21
shampoos, 54, 93
shark, 12, 151
shipping, 8, 23
shopping bags, 18
Sick Building Syndrome,
  170
silk, 97
sinus problems, 101
skin care products, 95
slaughterhouses, 10
sleep problems, 22

slow travel movement,
  143
slugging, 118
small growth, 117
smart growth
  communities, 117
soaps and detergents
  dishwashing, 60
  essential information, 54
  furniture cleaners, 61
  household cleaners, 59
  laundry, 64, 65
  school use, 170
socially responsible
  investing, 131
Sodexo, 176
sodium laureth
  sulfate, 64
sodium lauryl sulfate,
  64, 94
sodium polyacrylate, 161
soil testing, 79
solar energy, 32
  household appliances
    run by, 43
  off-grid living, 38
  system design, 43
soy foods, 17, 19
sperm damage, 35
spiders, 56
Sprig toys, 153
sprouts, 15
stain-resistant fabrics, 52
strawberries, 7
styrofoam, 71
sugar, 25
sunscreen, 95
sushi, 12
Sustainablebusiness.com,
  125
Sustainable Forestry
  Initiative, 166

Sustainable Travel
    International, 145
sweatshops, 130
swimming pools, 46
swordfish, 12, 151

t

Tai Chi, 99
tap water, 6
Tarte cosmetics, 92
tea, 22, 156
tea tree oil, 56
telephones, 68
television, 66
tempeh, 17
terpenes, 59
The Little Seed, 153
theobromine, 24
Thoreau, Henry David,
    150
thrift stores, 111, 130
tilefish, 12
titanium oxide, 95
tobacco products, 105
toilets, 57
toluene, 96
tooth care, 94
tourism, 145
toys, 153
trees
    for coffee production, 21
    for diaper production,
        161
    environmental benefits,
        110, 143
    for paper production, 62,
        69, 129
    planting campaigns, 110
    Sustainable Forestry
    Initiative, 166

wildfires, 180
    see also paper; rain forests
triclosan, 94
trifluralin, 153
Tris, 52
tuna, 12, 151
turkey, 72
Tutu, Desmond, 116

u

United Nations, 110
upcycling, 70
upholstery cleaner, 61
urban planning, 117, 125
U.S. Council of Mayors,
    114
U.S. Green Building
    Council, 126, 176
U.S. Public Interest
    Research Group,
    115

v

vaccinations, 160
valerian, 22
Van Gogh, Vincent, 45
vegan diet
    economic benefits, 13
    environmental impact, 9
    health benefits, 5, 14
Vegetarian Cats & Dogs
    (Peden), 63
vegetarian diet
    for children, 164
    economic benefits, 13
    environmental impact, 9
    health benefits, 5, 13
    for pets, 63

Victory Gardens, 113
vinegar, 59, 61, 65
vinyl products, 35
vitamin B12, 14, 15
vitamin D, 15, 100
vitamin E, 104
vitamins and supplements,
    104
volatile organic
    compounds, 170
    in household
        furnishings, 33, 41
    in paints, 55

w

Walden Pond, or Life in
    the Woods (Thoreau),
    150
wallpaper adhesive, 45
Ward, Barbara, 134
washing machine, 64
waste
    annual output, 69
    benefits of living simply,
        124
    composting, 78
    human, 57
    paper, 129
    pet, 63
    prescription drug
        disposal, 58, 96
    restaurant food, 9
    strategies to reduce, 38
water
    bathroom use, 57
    bottled, 6
    catchment systems, 36,
        43, 83
    in hydraulic fracking for
        natural gas, 43

water (*cont.*)
  laundry use of, 64
  prescription drug
    contamination, 58
  river pollutants, 96
  from septic systems, 57
  tap, 6
weatherizing and
    insulating, 37, 39, 125
web resources
  asbestos, 42
  brewpubs, 23
  carbon offsets, 144
  cell phone recyclers, 68
  children's toys, 153
  community activism,
    114
  community gardens, 113
  cool roof design, 36
  cosmetics database, 92
  Electronic Product
    Environmental
    Assessment Tool, 67
  geotourism, 145
  green grants, 126
  green restaurants, 9
  hybrid and electric cars,
    139
  investment counseling,
    131
  lead abatement, 45
  nursery furnishings, 153
  pet food recipes, 63
  product recycling, 69, 93

raw food menus, 15
Really, Really Free
  Market movement,
  112
seafood safety, 12
slow travel movement,
  143
sources for locally grown
  food, 8
urban environmental
  strategies, 114
urban gardening, 86
vegetarian lifestyle, 13
verifying environmental
  claims of companies, 55
water quality
  assessment, 6
wind power, 44
weddings, 179
Weleda, 94, 95
whooping cough, 160
wildfires, 180
windows
  cleaning, 59
  design, 37
wind power, 44
wine, 23
Winnie-the-Pooh, 76
witch hazel, 95
Women, Infants, and
  Children program,
  159
wood products
  baby toys, 153

cabinets and house
  furnishings, 52
flooring, 35
for food preparation, 26
forest sustainability, 143
outdoor decks, 47
playground equipment,
  166
wool, 97
worker's rights and
  working conditions, 10,
  18, 24
  in apparel industry, 97,
  130
workplace
  environmentalism, 123,
  127, 129
World Commission on
  Dams, 127
World Naked Bike Ride,
  142

y

yin/yang balance, 16
yoga, 98
Yoga Sutras of Patanjali,
  The, 98

z

zinc oxide, 95

about the author

SARA GILBERT is creator, executive producer, and co-host of CBS's Emmy-nominated *The Talk*. She first captured America's heart in her role as Darlene Conner in the long-running hit series *Roseanne*, for which she received two Emmy nominations. Since then she has appeared in movies and regularly on a variety of television shows, including *The Big Bang Theory, ER,* and *24.* A vegan mother of two and magna cum laude graduate of Yale, Gilbert has established herself in the Hollywood community and beyond as a trusted voice on family, health, parenting, and eco-conscious living.

about the type

This book was set in Scala, a typeface designed by Martin Majoor in 1991. It was originally designed for a music company in the Netherlands and then was published by the international type house FSI FontShop. Its distinctive extended serifs add to the articulation of the letterforms to make it a very readable typeface.